Interim Report

The Scientific Basis for Estimating Air Emissions from Animal Feeding Operations

Ad Hoc Committee on Air Emissions from Animal
Feeding Operations
Committee on Animal Nutrition
Board on Agriculture and Natural Resources
Board on Environmental Studies and Toxicology
Division on Earth and Life Studies

NATIONAL RESEARCH COUNCIL

NATIONAL ACADEMY PRESS
Washington, D.C.

NATIONAL ACADEMY PRESS · 2101 Constitution Avenue, NW ·
Washington, DC 20418

NOTICE: The project that is the subject of this report was approved by the Governing Board of the National Research Council, whose members are drawn from the councils of the National Academy of Sciences, the National Academy of Engineering, and the Institute of Medicine. The members of the committee responsible for the report were chosen for their special competences and with regard for appropriate balance.

This report has been reviewed by a group other than the authors according to procedures approved by a Report Review Committee consisting of members of the National Academy of Sciences, the National Academy of Engineering, and the Institute of Medicine.

This study was supported by Contract No. 68-D-01-69 between the National Academy of Sciences and the U.S. Environmental Protection Agency and Grant No. 59-0790-2-106 between the National Academy of Sciences and the U.S. Department of Agriculture. Any opinions, findings, conclusions, or recommendations expressed in this publication are those of the author(s) and do not necessarily reflect the views of the organizations or agencies that provided support for the project.

International Standard Book Number 0-309-08461-X

Additional copies of this report are available from National Academy Press, 2101 Constitution Avenue, N.W., Lockbox 285, Washington, DC 20055; (800) 624-6242 or (202) 334-3313 (in the Washington metropolitan area); Internet, http://www.nap.edu

Printed in the United States of America
Copyright 2002 by the National Academy of Sciences. All rights reserved.

THE NATIONAL ACADEMIES

National Academy of Sciences
National Academy of Engineering
Institute of Medicine
National Research Council

The **National Academy of Sciences** is a private, nonprofit, self-perpetuating society of distinguished scholars engaged in scientific and engineering research, dedicated to the furtherance of science and technology and to their use for the general welfare. Upon the authority of the charter granted it by the Congress in 1863, the Academy has a mandate that requires it to advise the federal government on scientific and technical matters. Dr. Bruce M. Alberts is president of the National Academy of Sciences.

The **National Academy of Engineering** was established in 1964, under the charter of the National Academy of Sciences, as a parallel organization of outstanding engineers. It is autonomous in its administration and in the selection of its members, sharing with the National Academy of Sciences the responsibility for advising the federal government. The National Academy of Engineering also sponsors engineering programs aimed at meeting national needs, encourages education and research, and recognizes the superior achievements of engineers. Dr. Wm. A. Wulf is president of the National Academy of Engineering.

The **Institute of Medicine** was established in 1970 by the National Academy of Sciences to secure the services of eminent members of appropriate professions in the examination of policy matters pertaining to the health of the public. The Institute acts under the responsibility given to the National Academy of Sciences by its congressional charter to be an adviser to the federal government and, upon its own initiative, to identify issues of medical care, research, and education. Dr. Kenneth I. Shine is president of the Institute of Medicine.

The **National Research Council** was organized by the National Academy of Sciences in 1916 to associate the broad community of science and technology with the Academy's purposes of furthering knowledge and advising the federal government. Functioning in accordance with general policies determined by the Academy, the Council has become the principal operating agency of both the National Academy of Sciences and the National Academy of Engineering in providing services to the government, the public, and the scientific and engineering communities. The Council is administered jointly by both Academies and the Institute of Medicine. Dr. Bruce M. Alberts and Dr. Wm. A. Wulf are chairman and vice chairman, respectively, of the National Research Council.

AD HOC COMMITTEE ON AIR EMISSIONS FROM ANIMAL FEEDING OPERATIONS

PERRY R. HAGENSTEIN (*Chair*), Institute for Forest Analysis, Planning, and Policy, Wayland, Massachusetts
ROBERT G. FLOCCHINI (*Vice-Chair*), University of California, Davis, California
JOHN C. BAILAR III, University of Chicago, Chicago, Illinois
CANDIS CLAIBORN, Washington State University, Pullman, Washington
RUSSELL R. DICKERSON, University of Maryland, College Park, Maryland
JAMES N. GALLOWAY, University of Virginia, Charlottesville, Virginia
MARGARET ROSSO GROSSMAN, University of Illinois at Urbana-Champaign, Urbana, Illinois
PRASAD KASIBHATLA, Duke University, Durham, North Carolina
RICHARD A. KOHN, University of Maryland, College Park, Maryland
MICHAEL P. LACY, University of Georgia, Athens, Georgia
CALVIN B. PARNELL, Jr., Texas A&M University, College Station, Texas
ROBBI H. PRITCHARD, South Dakota State University, Brookings, South Dakota
WAYNE P. ROBARGE, North Carolina State University, Raleigh, North Carolina
DANIEL A. WUBAH, James Madison University, Harrisonburg, Virginia
KELLY D. ZERING, North Carolina State University, Raleigh, North Carolina
RUIHONG ZHANG, University of California, Davis, California

Staff
JAMIE JONKER, Study Director
CHAD TOLMAN, Program Officer
TANJA PILZAK, Research Assistant
JULIE ANDREWS, Senior Project Assistant
STEPHANIE PADGHAM, Project Assistant
BRYAN SHIPLEY, Project Assistant

COMMITTEE ON ANIMAL NUTRITION

GARY L. CROMWELL (*Chair*), University of Kentucky, Lexington, Kentucky
C. ROSELINA ANGEL, University of Maryland, College Park, Maryland
JESSE P. GOFF, United States Department of Agriculture/Agricultural Research Service, Ames, Iowa
RONALD W. HARDY, University of Idaho, Hagerman, Idaho
KRISTEN A. JOHNSON, Washington State University, Pullman, Washington
BRIAN W. MCBRIDE, University of Guelph, Guelph, Ontario, Canada
KEITH E. RINEHART, Perdue Farms Incorporated, Salisbury, Maryland
L. LEE SOUTHERN, Louisiana State University, Baton Rouge, Louisiana
DONALD R. TOPLIFF, West Texas A&M University, Canyon, Texas

Staff
CHARLOTTE KIRK BAER, Program Director
JAMIE JONKER, Program Officer
STEPHANIE PADGHAM, Project Assistant

BOARD ON AGRICULTURE AND NATURAL RESOURCES

HARLEY W. MOON (*Chair*), Iowa State University, Ames, Iowa
CORNELIA B. FLORA, Iowa State University, Ames, Iowa
ROBERT B. FRIDLEY, University of California, Davis, California
BARBARA GLENN, Federation of Animal Science Societies, Bethesda, Maryland
LINDA GOLODNER, National Consumers League, Washington, D.C.
W.R. (REG) GOMES, University of California, Oakland, California
PERRY R. HAGENSTEIN, Institute for Forest Analysis, Planning, and Policy, Wayland, Massachusetts
GEORGE R. HALLBERG, The Cadmus Group, Inc., Waltham, Massachusetts
CALESTOUS JUMA, Harvard University, Cambridge, Massachusetts
GILBERT A. LEVEILLE, McNeil Consumer Healthcare, Denville, New Jersey
WHITNEY MACMILLAN, Cargill, Inc., Minneapolis, Minnesota
TERRY MEDLEY, DuPont Biosolutions Enterprise, Wilmington, Delaware
WILLIAM L. OGREN, U.S. Department of Agriculture (retired), Hilton Head, South Carolina
ALICE PELL, Cornell University, Ithaca, New York
NANCY J. RACHMAN, Novigen Sciences, Inc., Washington, D.C.
G. EDWARD SCHUH, University of Minnesota, Minneapolis, Minnesota
BRIAN STASKAWICZ, University of California, Berkeley, California
JOHN W. SUTTIE, University of Wisconsin, Madison, Wisconsin
JAMES TUMLINSON, USDA, ARS, Gainesville, Florida
JAMES J. ZUICHES, Washington State University, Pullman, Washington

Staff
CHARLOTTE KIRK BAER, Director
JULIE ANDREWS, Senior Project Assistant

BOARD ON ENVIRONMENTAL STUDIES AND TOXICOLOGY

GORDON ORIANS *(Chair)*, University of Washington, Seattle, Washington
JOHN DOULL *(Vice Chair)*, University of Kansas Medical Center, Kansas City, Kansas
DAVID ALLEN, University of Texas, Austin, Texas
INGRID C. BURKE, Colorado State University, Fort Collins, Colorado
THOMAS BURKE, Johns Hopkins University, Baltimore, Maryland
WILLIAM L. CHAMEIDES, Georgia Institute of Technology, Atlanta, Georgia
CHRISTOPHER B. FIELD, Carnegie Institute of Washington, Stanford, California
DANIEL S. GREENBAUM, Health Effects Institute, Cambridge, Massachusetts
BRUCE D. HAMMOCK, University of California, Davis, California
ROGENE HENDERSON, Lovelace Respiratory Research Institute, Albuquerque, New Mexico
CAROL HENRY, American Chemistry Council, Arlington, Virginia
ROBERT HUGGETT, Michigan State University, East Lansing, Michigan
JAMES H. JOHNSON, Howard University, Washington, D.C.
JAMES F. KITCHELL, University of Wisconsin, Madison, Wisconsin
DANIEL KREWSKI, University of Ottawa, Ottawa, Ontario, Canada
JAMES A. MACMAHON, Utah State University, Logan, Utah
WILLEM F. PASSCHIER, Health Council of the Netherlands, The Hague, The Netherlands
ANN POWERS, Pace University School of Law, White Plains, New York
LOUISE M. RYAN, Harvard University, Boston, Massachusetts
KIRK SMITH, University of California, Berkeley, California
LISA SPEER, Natural Resources Defense Council, New York, New York

Staff
JAMES J. REISA, Director
RAY WASSEL, Program Director
MIMI ANDERSON, Senior Project Assistant

Acknowledgments

This report represents the integrated efforts of many individuals. The committee thanks all those who shared their insights and knowledge to bring the document to fruition. We also thank all those who provided information at our public meetings and who participated in our public sessions.

During the course of its deliberations, the committee sought assistance from several people who gave generously of their time to provide advice and information that were considered in its deliberations. Special thanks are due the following:

BOB BOTTCHER, North Carolina State University, Raleigh, North Carolina
GARTH BOYD, Murphy-Brown LLC, Warsaw, North Carolina
LEONARD BULL, Animal and Poultry Waste Center, Raleigh, North Carolina
TOM CHRISTENSEN, United States Department of Agriculture, Beltsville, Maryland
JOHN D. CRENSHAW, Eastern Research Group, Research Triangle Park, North Carolina
TONY DELANY, National Center for Atmospheric Research, Boulder, Colorado
ERIC GONDER, Goldsboro Milling Company, Goldsboro, North Carolina
ALEX GUENTHER, National Center for Atmospheric Research, Boulder, Colorado

LOWRY HARPER, United States Department of Agriculture, Watkinsville, Georgia
BRUCE HARRIS, United States Environmental Protection Agency, Research Triangle Park, North Carolina
TOM HORST, National Center for Atmospheric Research, Boulder, Colorado
DONALD JOHNSON, Colorado State University, Fort Collins, Colorado
RENEE JOHNSON, United States Environmental Protection Agency, DC
JOHN H. MARTIN, Jr., Hall Associates, Dover, Delaware
F. ROBERT MCGREGOR, Water and Waste Engineering, Inc., Denver, Colorado
BOB MOSER, ConAgra Beef, Kersey, Colorado
DANIEL MURPHY, National Oceanographic and Atmospheric Administration, Boulder, Colorado
ROY OOMMEN, Eastern Research Group, Research Triangle Park, North Carolina
JOSEPH RUDEK, Environmental Defense, Raleigh, North Carolina
GARY SAUNDERS, North Carolina Department of Environment and Natural Resources, Raleigh, North Carolina
SUSAN SCHIFFMAN, Duke University, Durham, North Carolina
SALLY SHAVER, United States Environmental Protection Agency, Research Triangle Park, North Carolina
MARK SOBSEY, University of North Carolina, Chapel Hill, North Carolina
JOHN SWEETEN, Texas A&M University, Amarillo, Texas
RANDY WAITE, United States Environmental Protection Agency, Research Triangle Park, North Carolina
JOHN T. WALKER, United States Environmental Protection Agency, Research Triangle Park, North Carolina

The committee is grateful to members of the National Research Council (NRC) staff who worked diligently to maintain progress and quality in its work.

This report has been reviewed in draft form by individuals chosen for their diverse perspectives and technical expertise, in accordance with procedures approved by the National Research Council's Report Review Committee. The purpose of this independent review is to provide candid and critical comments that will assist the institution in making its published report as sound as possible and to ensure that the report meets institutional standards for objectivity, evidence, and responsiveness to the study charge. The review comments and draft manuscript remain confidential to protect the integrity of the deliberative process. We wish to thank the following individuals for their review of this report:

DAVID T. ALLEN, The University of Texas, Austin, Texas
VAN C. BOWERSOX, Illinois State Water Survey, Champaign, Illinois
ELLIS B. COWLING, North Carolina State University, Raleigh, North Carolina
ALBERT J. HEBER, Purdue University, West Lafayette, Indiana
JAMES A. MERCHANT, The University of Iowa, Iowa City, Iowa
DEANNE MEYER, University of California, Davis, California
ROGER A. PIELKE, Colorado State University, Fort Collins, Colorado
WENDY J. POWERS, Iowa State University, Ames, Iowa
JOSEPH RUDEK, Environmental Defense, Raleigh, North Carolina
JAMES J. SCHAUER, University of Wisconsin, Madison, Wisconsin
ANDREW F. SEIDL, Colorado State University, Fort Collins, Colorado

Although the reviewers listed above have provided many constructive comments and suggestions, they were not asked to endorse the conclusions or recommendations nor did they see the final draft of the report before its release. The review of this report was overseen by Thomas Graedel, Yale University, New Haven, Connecticut and May Berenbaum, University of Illinois, Champaign, Illinois. Appointed by the National Research Council, they were responsible for making certain that an independent examination of this report was carried out in accordance with institutional procedures and that all review comments were carefully considered. Responsibility for the final content of this report rests entirely with the authoring committee and the institution.

Preface

This is an interim report of the *ad hoc* Committee on Air Emissions from Animal Feeding Operations of the National Research Council's Committee on Animal Nutrition. A final report is expected to be issued by the end of 2002. The interim report is intended to provide the committee's findings to date on assessment of the scientific issues involved in estimating air emissions from individual animal feeding operations (swine, beef, dairy, and poultry) as related to current animal production systems and practices in the United States. The committee's final report will include an additional assessment within eight broad categories: industry size and structure, emission measurement methodology, mitigation technology and best management plans, short- and long-term research priorities, alternative approaches for estimating emissions, human health and environmental impacts, economic analyses, and other potential air emissions of concern.

This interim report focuses on identifying the scientific criteria needed to ensure that estimates of air emission rates are accurate, the basis for these criteria in the scientific literature, and uncertainties associated with them. It also includes an assessment of the emission-estimating approaches in a recent U.S. Environmental Protection Agency (EPA) report *Air Emissions from Animal Feeding Operations* (EPA, 2001a). Finally, it identifies economic criteria needed to assess emission mitigation techniques and best management practices.

The committee held three meetings in preparing this interim report and developing material for its final report. People knowledgeable about air emissions issues, including representatives of EPA, the U.S. Department of

Agriculture (USDA), academia, the animal feeding industry, and the public, presented relevant information at each of the meetings, which were held in Washington, D.C., Durham, North Carolina, and Denver, Colorado. Field visits to animal feeding operations were also conducted. The committee also reviewed various peer-reviewed and non-peer-reviewed literature describing the issues, the science that lies behind methods for measuring and estimating emissions, and materials prepared by and for EPA and USDA.

The committee relied on the expertise and knowledge of its members, who represent a range of disciplinary backgrounds, including epidemiology and biostatistics, environmental engineering, atmospheric and tropospheric chemistry, biogeochemistry, environmental sciences, agricultural law, animal nutrition, agricultural engineering, soils and physical chemistry, microbiology, agricultural and resource economics, emission measurement and characterization, and biological engineering.

Perry Hagenstein, Chair
Robert Flocchini, Vice-Chair
Committee on Air Emissions
from Animal Feeding Operations

Contents

EXECUTIVE SUMMARY   1

1. INTRODUCTION   7
   Animal Production, 13
   Emissions from Animal Feeding Operations, 14
   Distribution of Emitted Pollutants, 20

2. DETERMINING EMISSION FACTORS   23
   Introduction, 23
   Scientific Criteria, 25
   Published Literature, 29
   Characterizing Variability, 43
   Statistical Uncertainty, 49

3. MODELS FOR ESTIMATING EMISSIONS   54
   EPA Model Farm Construct, 54
   Industry Characterization, 57
   Process-Based Model Farm Approach, 58
   Mitigation Technologies and Best Management Practices, 61

4. ASSESSING THE EFFECTIVENESS OF EMISSION MITIGATION
   TECHNIQUES AND BEST MANAGEMENT PRACTICES   66
   Criteria for Evaluating Emissions Effects of Mitigation Techniques, 67
   Criteria for Evaluating Economic Effects of Mitigation Techniques, 68
   Partial Budgeting/Selected Cost and Returns Estimation, 72
   Other Considerations in Evaluation of Mitigation Techniques, 73

TABLES AND FIGURES
Tables
1-1 Current Hydrogen Sulfide Standards in Various States, 16
1-2 National Air Quality Standards for Particulate Matter, 19
1-3 Typical Lifetimes in the Planetary Boundary Layer for Pollutants Emitted from Animal Feeding Operations, 21
2-1 Odor Emission Rates from Animal Housing as Reported in the Literature, 43
2-2 Calculated Emission Rates of Ammonia from Primary Anaerobic Swine Lagoons as a Function of Measurement Method and Measurement Period, 50
3-1 Classification of Emissions by Likely Intended Use of Emission Factors, 57

Figures
2-1 Relative excretion rate of nitrogen versus day in the life cycle of a grow-finish hog at a commerical swine production facility in the southeastern United States, 47
3-1 A process-based model of emissions from an animal feeding operation, 59

REFERENCES   75

APPENDIXES
   A. Statement of Task   89
   B. Glossary   90
   C. Public Meeting Agendas   93
   D. Twenty-three Model Farms Described by EPA   98
   E. About the Authors   101

Executive Summary

Concern with possible environmental and health effects of air emissions generated from animal feeding operations (AFOs) has grown with the increasing size, geographic concentration, and suburbanization of these operations in what was formerly rural, sparsely populated agricultural land. This interim report, prepared at the request of the Environmental Protection Agency (EPA), evaluates the current knowledge base and approaches for estimating air emissions from AFOs. The issues regarding emissions from AFOs are much broader than the interests of any one federal agency. In recognition of this, the U.S. Department of Agriculture (USDA) joined EPA in the request for this study.

Generating reasonably accurate estimates of air emissions from AFOs is difficult. The operating environment for these farms is complex. The species of animals are varied (e.g., swine, beef and dairy cattle, poultry), and farm practices differ not only between species, but also among farms for each species. The operations vary in size (this report is concerned with AFOs as defined by EPA; see Appendix B) and differ by region across the country. The chemical composition of the emissions varies depending on animal species, feeding regimes and practices, manure management practices, and the way in which the animals are housed. Much of the air emissions come from the storage and disposal of the manure (the term here is used to mean both urine and feces, and may also include litter or bedding materials) that is part of every AFO, but some also comes from dust produced by the handling of feed and the movement of animals on manure, as well as from the animals themselves. Meteorologic conditions, of course, are an important factor. Estimates of emission rates generated in one type of AFO may not translate readily into others.

EPA has a variety of needs for accurate estimation of air emissions from AFOs. Increasing pressure has been placed on the agency to address these emissions through the Clean Air Act and other federal regulations, and EPA has indicated the need to do so in the future. Also pressing, EPA is under court order to establish new water quality rules by December 2002. The current study will focus on ways to estimate these emissions prior to December 2002 to additionally help assure that rules aimed at improving water quality do not have negative impacts on air emissions.

This interim report is intended to provide findings to date on a series of specific questions from EPA regarding the following general issues: identifying the scientific criteria needed to ensure that estimates of air emission rates are accurate, the basis for these criteria in the scientific literature, and the uncertainties associated with them. It also includes an assessment of the emission estimating approaches in a recent report *Air Emissions From Animal Feeding Operations* (EPA, 2001a). Finally, it identifies economic criteria needed to assess emission mitigation techniques and best management practices. The committee has answered the following sets of questions in the interim report within the confines of the Statement of Task (see Appendix A):

- What are the scientific criteria needed to ensure that reasonably appropriate estimates of emissions are obtained? What are the strengths, weaknesses, and gaps of published methods to measure specific emissions and develop emission factors that are published in the scientific literature? How should the variability due to regional differences, daily and seasonal changes, animal life stage, and different management approaches be characterized? How should the statistical uncertainty in emissions measurements and emissions factors be characterized in the scientific literature?

- Are the emission estimation approaches described in the EPA report *Air Emissions from Animal Feeding Operations* (EPA, 2001a) appropriate? If not, how should industry characteristics and emission mitigation techniques be characterized? Should model farms be used to represent the industry? If so, how? What substances should be characterized and how can inherent fluctuations be accounted for? What components of manure should be included in the estimation approaches (e.g., nitrogen, sulfur, volatile solids [see Appendix B])? What additional emission mitigation technologies and management practices should be considered?

- What criteria, including capital costs, operating costs, and technical feasibility, are needed to develop and assess the effectiveness of emission mitigation techniques and best management practices?

EXECUTIVE SUMMARY 3

The goal of EPA (2001a) was to "develop a method for estimating emissions at the individual farm level." To accomplish this, EPA (2001a) developed a set of 23 model farms (see Appendix D) intended to represent the majority of commercial-scale AFOs. Each model farm included three variable elements: a confinement area, manure management system, and land application method. The manure management system was subdivided into solid separation and manure storage activities.

Given the specific nature of the questions answered, the committee has not yet addressed some of the broader issues related to AFOs. To the extent possible, these will be addressed in its final report, which will build on the findings of this interim report and include a more detailed response to the committee's full Statement of Task (see Appendix A). The need for further discussion of some issues in the final report is indicated in various places in this report. These issues fall in eight broad categories: (1) industry size and structure, (2) emission measurement methodology, (3) mitigation technology and best management plans, (4) short- and long-term research priorities, (5) alternative approaches for estimating emissions, (6) human health and environmental impacts, (7) economic analyses, and (8) other potential air emissions of concern.

This interim report represents the consensus views of the committee and has been formally reviewed in accordance with National Research Council (NRC) procedures. In answering these questions and addressing its Statement of Task (Appendix A), the committee has come to consensus on eight findings for the interim report. The basis of these findings is discussed more extensively in the body of the report.

Finding 1: Proposed EPA regulations aimed at improving water quality may affect rates and distributions of air emissions from animal feeding operations.
Discussion: Regulations aimed at protecting water quality would probably affect manure management at the farm level, especially since they might affect the use of lagoons and the application of manure on cropland or forests. For example, the proposed water regulations may mandate nitrogen (N) or phosphorus (P) based comprehensive nutrient management plans (CNMPs). AFOs could be limited in the amount of manure nitrogen and phosphorus that could be applied to cropland. If there is a low risk of phosphorus runoff as determined by a site analysis, farmers will be permitted to overapply phosphorous. However, they will still be prohibited from applying more nitrogen than recommended for crop production. Many AFOs (those currently without CNMPs) likely will have more manure than they can use on their own cropland, and manure export may be cost prohibitive. Thus, AFOs will have an incentive to use crops and management practices that employ applied nitrogen inefficiently (i.e., volatilize ammonia) to decrease the nitrogen remaining after storage or increase the nitrogen requirement for crop production. These

practices may increase nitrogen volatilization to the air. The committee was not informed of specific regulatory actions being considered by EPA (beyond those addressed in the *Federal Register*) to meet its December 2002 deadline for proposing regulations under the Clean Water Act.

Finding 2: In order to understand health and environmental impacts on a variety of spatial scales, estimates of air emissions from AFOs at the individual farm level, and their dependence on management practices, are needed to characterize annual emission inventories for some pollutants and transient downwind spatial distributions and concentrations for others.

Discussion: Management practices (e.g., feeding, manure management, crop management) vary widely among individual farms. Estimates of emissions based on regional or other averages are unlikely to capture significant differences among farms that will be relevant for guiding emissions management practices aimed at decreasing their effects. Information on the spatial relationships among individual farms and the dispersion of air emissions from them is needed. Furthermore, developing methods to estimate emissions at the individual farm level was the stated objective of EPA's recent study (EPA, 2001a).

Finding 3: Direct measurements of air emissions at all AFOs are not feasible. Nevertheless, measurements on a statistically representative subset of AFOs are needed and will require additional resources to conduct.

Discussion: Although it is possible in a carefully designed research project to measure concentrations and airflows (e.g., building ventilation rates) to estimate air emissions and attribute them to individual AFOs, it is not practical to conduct such projects for more than a small fraction of AFOs. Direct measurements for sample farms will be needed in research programs designed to develop estimates of air emissions applicable to various situations.

Finding 4: Characterizing feeding operations in terms of their components (e.g., model farms) may be a plausible approach for developing estimates of air emissions from individual farms or regions as long as the components or factors chosen to characterize the feeding operation are appropriate. The method may not be useful for estimating acute health effects, which normally depend on human exposure to some concentration of toxic or infectious substance for short periods of time.

Discussion: The components or factors used to characterize feeding operations are chosen for their usefulness in explaining dependent variables, such as the mass of air emissions per unit of time. The emission factor method, which is based on the average amount of an emitted substance per unit of activity per year (e.g., metric tons of ammonia per thousand head of cattle per year), can be useful in estimating annual regional emissions inventories for some pollutants, provided that sufficient data of adequate quality are available for estimating the relationships.

Finding 5: Reasonably accurate estimates of air emissions from AFOs at the individual farm level require defined relationships between air emissions and various factors. Depending on the character of the AFOs in question, these factors may include animal types, nutrient inputs, manure handling practices, output of animal products, management of feeding operations, confinement conditions, physical characteristics of the site, and climate and weather conditions.
Discussion: The choice of independent variables used to make estimates of air emissions from AFOs will depend on the ability of the variables to account for variations in the estimates and on the degree of accuracy desired, based on valid measurements at the farm level. Past research indicates that some combination of the indicated variables is likely to be important for estimates of air emissions for the kinds of operations considered in this report. The specific choices will depend on the strength of the relationships for each kind of emission and each set of independent variables.

Finding 6: The model farm construct as described by EPA (2001a) cannot be supported because of weaknesses in the data needed to implement it.
Discussion: Of the nearly 500 possible literature sources for estimating emissions factors identified for EPA (2001a), only 33 were found by the report's authors to be suitable for use in the model farm construct. The committee judged them to be insufficient for the intended use. The breadth in terms of kinds of animals, management practices, and geography in this model farm construct suggests that finding adequate information to define emission factors is unlikely to be fruitful at this time.

Finding 7: The model farm construct used by EPA (2001a) cannot be supported for estimating either the annual amounts or the temporal distributions of air emissions on an individual farm, subregional, or regional basis because the way in which it characterizes feeding operations is inadequate.
Discussion: Variations in many factors that could affect the annual amounts and temporal patterns of emissions from an individual AFO are not adequately considered by the EPA (2001a) model farm construct. The potential influences of geographic (e.g. topography and land use) and climatic differences, daily and seasonal weather cycles, animal life stages, management approaches (including manure management practices and feeding regimes), and differences in state regulations are not adequately considered. Furthermore, aggregating emissions from individual AFOs using the EPA (2001a; not a stated objective) model farm construct for subregional or regional estimates cannot be supported for similar

reasons. However, with the appropriate data identified there may be viable alternatives to the currently proposed approach.

Finding 8: A process-based model farm approach that incorporates "mass balance" constraints for some of the emitted substances of concern, in conjunction with estimated emission factors for other substances, may be a useful alternative to the model farm construct defined by EPA (2001a). The committee plans to explore issues associated with these two approaches more fully in its final report.
Discussion: The mass balance approach, like EPA's model farm approach, starts with defining feeding operations in terms of major stages or activities. However, it focuses on those activities that determine the movement of nutrients and other substances into, through, and out of the system. Experimental data and mathematical modeling are used to simulate the system and the movement of reactants and products through each component of the farm enterprise. In this approach, emissions of elements (such as nitrogen) cannot exceed their flows into the system.

1

Introduction

This interim report provides the U.S. Environmental Protection Agency (EPA), the U.S. Department of Agriculture (USDA), other federal and state agencies, the animal feeding industry, and the general public an initial assessment of the methods and quality of data used in estimating air emissions from animal feeding operations (AFOs as defined by EPA; see Appendix B). These emissions, their impacts, and the methods used to mitigate them affect the health and well-being of individual farms, the agricultural economy, the associated environments, and people. The scientific aspects of this broad issue deserve attention, both in the near term as possible revisions of federal water quality regulations are being considered, and in the longer term as attention shifts to ways to mitigate air emissions.

The stakes in this issue are large. More and more livestock are raised for at least part of their lives in AFOs in response to economic factors that encourage further concentration. The impacts on the air in surrounding areas have grown to a point where further actions to mitigate them appear likely. The overall study, of which this interim report is part, has been requested to help ensure that choices among alternatives are made on the basis of information that meets the tests of scientific accuracy.

The committee has been sensitive to the fact that its findings are not being written on a blank slate. The types of actions that might ultimately result from this and other reports could include various kinds of regulation, public incentive approaches, and technical assistance, all of which are already being used to some extent by the states and federal agencies. The committee also notes that this interim report will be supplemented by a final report in another six months, and that some of the discussions of possible approaches to estimating air emissions are being left for that report as noted in relevant places in this interim report. The committee has answered the following sets of questions in the interim report within the confines of the Statement of Task (see Appendix A):

- What are the scientific criteria needed to ensure that reasonably appropriate estimates of emissions are obtained? What are the strengths, weaknesses, and gaps of published methods to measure specific emissions and develop emission factors that are published in the scientific literature? How should the variability due to regional differences, daily and seasonal changes, animal life stage, and different management approaches be characterized? How should the statistical uncertainty in emissions measurements and emissions factors be characterized in the scientific literature?
- Are the emission estimation approaches described in the EPA report *Air Emissions from Animal Feeding Operations* (EPA, 2001a) appropriate? If not, how should industry characteristics and emission mitigation techniques be characterized? Should model farms be used to represent the industry? If so, how? What substances should be characterized and how can inherent fluctuations be accounted for? What components of manure should be included in the estimation approaches (e.g., nitrogen, sulfur, volatile solids [see Appendix B])? What additional emission mitigation technologies and management practices should be considered?
- What criteria, including capital costs, operating costs, and technical feasibility, are needed to develop and assess the effectiveness of emission mitigation techniques and best management practices?

Given the specific nature of the questions posed by EPA, the committee has not yet addressed some of the longer-term issues related to AFOs. To the extent possible, these will be addressed in the final report, which will build upon the findings of this interim report and include a more detailed response to the committee's full Statement of Task (see Appendix A). The need for further discussion in the final report is indicated for some specific concerns in various places in this report. The topics to be covered in the final report fall in eight broad categories: (1) industry size and structure, (2) emission measurement methodology, (3) mitigation technology and best management plans, (4) short- and long-term research priorities, (5) model farm approaches, (6) human health and environmental impacts, (7) economic analyses, and (8) other potential air emissions of concern.

The quality of data for estimating air emissions from AFOs is an issue throughout this report. The committee's inclination at first was to refer only to data from peer-reviewed sources. It soon became evident that this would eliminate a number of references that were prepared and relied upon by federal and state agencies, including the EPA (2001a) report that the committee is

INTRODUCTION

directed to review as part of its assignment. These reports sometimes rely on information from primary sources that have been peer reviewed, in which case they would meet the standard generally adopted by the committee. The committee decided that it would use results presented in these non-peer-reviewed or "gray literature" reports as long as it could determine that they reflected peer-reviewed sources. It also decided that it would clearly indicate instances where it believed that judicious use of non-peer-reviewed reports was needed.

EPA may use information from this project in determining how it will approach regulating both air and water quality impacts of AFOs. Substantial emissions of nitrogen (N), sulfur (S), carbon (C), particulate matter (PM), and other substances from AFOs do occur and cannot be ignored. This interim report also makes reference to possible influences that regulations proposed by the EPA Office of Water may have on aggravating air emissions from AFOs. The EPA's Office of Air and Radiation's concern with the possible effect of water quality regulations on air emissions is well placed. Effects on air emissions of nutrient management practices currently recommended to protect water quality are generally unknown. In addition to potential conflicts between air quality and regulations aimed at improving water quality, state regulations based on inadequate air emissions information may lead to inappropriate actions. Better understanding of the reliability of air emissions estimates will help EPA and the states to assess the appropriateness of regulations.

The potential effects on air emissions from changes in water quality regulations for AFOs will be difficult to predict, especially given the large number of AFOs in existence and the substantial number of animals involved. Changes induced through new water quality regulations could be either positive or negative in their effects on air quality. For example, the proposed water regulations may mandate nitrogen and phosphorus based comprehensive nutrient management plans (CNMPs). AFOs could be limited in the amount of manure nitrogen and phosphorus that could be applied to cropland. If there is a low risk of phosphorus runoff as determined by a site analysis, farmers may be permitted to overapply phosphorus. However, they will still be prohibited from applying more nitrogen than recommended for crop production. Many AFOs (those currently without CNMPs) likely will have more manure than they can use on their own cropland, and manure export may be cost prohibitive. Thus, AFOs will have an incentive to use crops and management practices that employ applied nitrogen inefficiently (i.e., volatilize ammonia) to decrease the nitrogen remaining after storage or increase the requirement for nitrogen on crop production. These practices could increase nitrogen volatilization to the air. AFOs with limited space to apply manure to fertilize their crops would have to adopt alternative management practices. Effects on air emissions of dispersal of manure across additional cropland (if available) must be considered. Although the transport of manure off-site reduces the emissions associated with that AFO, it does not guarantee an overall reduction of emissions into the environment. The committee recognizes that the EPA Office of Air and Radiation, and Office of Water face a considerable task in drafting new regulations and evaluating proposed regulations in terms of their relative impacts on air and water quality.

> **Finding 1: Proposed EPA regulations aimed at improving water quality may affect rates and distributions of air emissions from animal feeding operations.**

Regulations developed by the EPA's Office of Air and Radiation for AFOs will be influenced in part by existing National Ambient Air Quality Standards (NAAQS; EPA, 2002). These standards define concentration limits for ambient concentrations of six criteria pollutants (carbon monoxie, nitrogen dioxide, ozone, lead, PM10, and sulfur dioxide) based on health effects. Exceedances of these standards can result in areas being classified as "nonattainment" areas. The state implementation plans (SIPs) subsequently approved by EPA are plans for bringing these areas into attainment. SIPs may include sources of pollutants targeted for reduction. These are usually regulated by decreasing the allowable emission rates established by the permit control at each source needed to meet the NAAQS. States can legislate more stringent ambient air quality standards within their boundaries. Several of the substances emitted from AFOs that are of concern in this report are not regulated under NAAQS; examples include ammonia, hydrogen sulfide, and odor.

Developing SIPs for a region that contains AFOs may require knowledge of their air emissions. AFOs can differ significantly from each other in terms of construction, management, and operation. They can be widely distributed across the landscape or concentrated in geographic regions. To be effective, regulatory actions must ultimately account for emissions at the individual farm level and be based on information that can be used to attribute emissions to specific operations. Estimates of emissions at the state or regional level (e.g., across a watershed or river basin) may be sufficient to trigger the need for regulatory action. However, such actions, if needed, will ultimately depend on the ability to assign emissions to the individual operations that produce them. Application of remediation policies will in turn require knowledge of emissions from the individual components of AFOs.

> **Finding 2: In order to understand health and environmental impacts on a variety of spatial scales, estimates of air emissions from AFOs at the individual farm level, and their dependence on management practices, are needed to characterize annual emission inventories for some pollutants and transient downwind spatial distributions and concentrations for others.**

Estimating emissions of gases, PM, and other substances from AFOs is technically difficult. The variety of emissions; the different conditions under which they are emitted; the subsequent mixing, chemical reactions, and deposition following emission; the types and sizes of emitting operations; and the difficulty of obtaining representative samples all contribute to the challenge of accurately characterizing AFOs as emission sources. As reflected by EPA (2001a), an attempt was made to address the need for emissions estimates from individual AFOs (Finding 2) and to address the difficulty in characterizing AFOs as emissions sources by developing the concept of model farms. By judicious selection of criteria, emission factors obtained from the scientific literature for components of those model farms may allow for calculation of the desired estimate of annual mass emissions from a single AFO. To that end, the quality and lack of these data are discussed in detail in Chapter 2. The only remaining requirements would be assigning an individual AFO to a specific model farm category and an accounting of the animal units (AUs as defined by EPA and used throughout this report; see Appendix B) housed there. The approach outlined by EPA (2001a) could be interpreted as representing a compromise between the physical impracticality of installing monitoring equipment on every AFO (due to cost and the lack of standardized emission measurement methodologies that can be adopted for routine monitoring) and the growing public pressure to consider rural air quality as an integral part of resource management.

The committee supports the proposition that it is impractical to consider installing monitoring equipment at every AFO. First, emissions from AFOs are not typical of point sources since there are usually few convenient centrally located points from which to monitor emissions. Second, determining source emissions from AFOs should not be confused with monitoring atmospheric concentrations of gases, PM, or other substances. Measurement of atmospheric concentrations of substances is an important component in determining emissions, but application of meteorological models with other complementary data are often necessary to back-calculate emission rates or fluxes for gases and PM. In addition, no standard methods have been developed for measuring source emissions that state agencies could adopt for monitoring individual operations, let alone advising individuals on deployment and measurement strategies, given the diversity in design and operation of AFOs. Routine monitoring of air quality is employed for compliance purposes in many industries (i.e., electrical power, automobiles); however such efforts are based on many years of research to develop models to predict the emission from these anthropogenic sources with some degree of confidence. A corresponding investment of time and resources has not been made in understanding emissions from biological systems such as AFOs; however these research measurements are sorely needed.

Finding 3: Direct measurements of air emissions at all AFOs are not feasible. Nevertheless, measurements on a statistically representative subset of AFOs are needed and will require additional resources to conduct.

The committee also agrees that characterizing AFOs in terms of their production components (e.g., model farms) may in general be a plausible approach for developing estimates of air emissions. EPA (2001a) developed a set of 23 model farms (see Appendix D) intended to represent the majority of commercial-scale AFOs. Each model farm included three variable elements: a confinement area, manure management system, and land application method. The manure management system was subdivided into solid separation and manure storage activities.

A number of arguments exist to support an approach such as that outlined by EPA (2001a) with the creation of model farms. Most AFOs can be subdivided according to different manure management systems that are in turn constructed of individual processing steps. Animal housing units are often of a specified design depending on animal age and type. Although housing units may vary in design among farms, within an individual farm the housing units are generally uniform with respect to size, ventilation, and number of days animals are kept in each house. Feed formulations are also generally controlled uniformly as a function of animal age and stage of production. Animal growth across its life is often predicted through the use of models. Variations in ambient temperature due to seasonal changes no doubt cause changes in housing emissions due to the need to increase or decrease ventilation to remove or conserve heat. Ventilation protocols designed to control temperature and humidity may help to decrease concentrations of air emissions and maintain animal health. Thus, on a yearly basis, it may be possible to account for these seasonal variations. It could be argued that expressing emissions on a yearly basis would also tend to average out rotations of animals in and out of housing units; animal age varies between housing units on many AFOs.

Emissions of gases such as ammonia (NH_3) from manure treatment lagoons are dictated to a large extent by the ambient air temperature (through its influence on lagoon water temperature), lagoon pH, wind speed across the lagoon, and dissolved ammonium ion (NH_4^+) concentration and are relatively independent of week-to-week variations in loading of animal manure. Changes in NH_3 emissions due to changes in ambient temperature could conceivably be accounted for through the generation of regression models relating temperature, pH, and dissolved ammonium ion concentration. Similar examples could be given for other types of manure management systems; it is reasonable to assume that individual processing steps within a given manure management system could be characterized by single emission factors that when combined, would lead to a viable estimate of emissions for each type of model farm. The only limitation in the approach is the lack of accurate emission factors based on field data for the individual processing steps and interactions among these steps.

In opposition to the above statements are the intuitive arguments that AFOs are complicated systems with inherent variability because of differences in physical design and the fact that AFOs are biological systems with daily, seasonal, and probably yearly cycles. The biological complexity of AFOs exists at both the macro- and the microscales. The macroscale may include the various growth stages of animals being produced, with changes in feed formulation, consumption, productivity, and manure produced. The microscale may include microbial activity within the animal and in excreted animal manure; all microbial processes depend to some degree on changes in temperature, oxygen concentrations, and moisture content. Measured emission rates will necessarily have a component of uncertainty that will carry over to emission factors generated from them. Deriving an estimate of this uncertainty is necessary in order to compare estimated emissions among individual AFOs and to compare the emissions from a single AFO to regulatory limits.

A substantial body of research shows that the air emissions from AFOs depend on a variety of factors that vary among the different kinds of operations. It is reasonable to expect that there are particular sets of factors, to be established with statistical techniques, that will be most useful in estimating air emissions for each kind of operation. However, the committee believes that the model farm construct currently outlined (EPA, 2001a) has not identified all of the factors necessary to characterize emissions from individual AFOs.

Finding 4: Characterizing feeding operations in terms of their components (e.g., model farms) may be a plausible approach for developing estimates of air emissions from individual farms or regions as long as the components or factors chosen to characterize the feeding operation are appropriate. The method may not be useful for estimating acute health effects, which normally depend on human exposure to some concentration of toxic or infectious substance for short periods of time.

ANIMAL PRODUCTION

In 1995, at any given time there were approximately 13 billion chickens, 1.3 billion cattle, and 0.9 billion pigs worldwide; of these, 1.6 billion chickens, 0.1 billion cattle, and 0.06 billion pigs were located in the United States (Food and Agriculture Organization, 2002). The U.S. stocks sustained the production of 11.5 Tg of chicken meat, 11.6 Tg of beef and veal, and 8.1 Tg of pork. These products are important sources of calories and protein; in 1993, they supplied 28 percent of the calories and 64 percent of the protein consumed by humans in the United States (Council for Agricultural Science and Technology, 1999). In addition to producing food, animals also produce waste. In 1997, 1 x 10^{12} kg (10^3 Tg) of manure was excreted in the United States, with confined animals producing about 40 percent of it (Kellogg et al., 2000).

This report addresses the issue of air emissions from AFOs with a special focus on the gases ammonia (NH_3), nitric oxide (NO), hydrogen sulfide (H_2S), nitrous oxide (N_2O), and methane (CH_4); the general class of materials designated volatile organic compounds (VOCs); odor-causing compounds; and the aerosol classes PM2.5 and PM10 (particulate matter having aerodynamic diameters less than 2.5 and less than 10 micrometers (μm), respectively). In the remaining sections estimates of global emissions are presented based on reviews from a number of sources, often using emission factors. Given the uncertainties in emission factors, these global emissions also have uncertainties, which are limited by constraints on global budget terms (such as loss rates). Estimates of aggregated emissions rates from all sources can be at least partially validated by measurement of spatial and temporal differences in ambient air concentrations. Accuracy of attribution of total emissions to individual sources is limited by incomplete lists of the sources, and errors in assumed emission factors for each source. The source-specific estimates provided in the following sections are subject to these limitations but are presented to give the reader a general sense of each source's importance.

EMISSIONS FROM ANIMAL FEEDING OPERATIONS

Ammonia

The nitrogen in animal manure can be converted to ammonia by a combination of mineralization, hydrolysis, and volatilization (Oenema et al., 2001). On a global scale, animal farming systems emit to the atmosphere ~20 Tg N/yr as NH_3 (Galloway and Cowling, 2002), about 65 percent of total NH_3 emissions from terrestrial systems (van Aardenne et al., 2001). In the United States, about 6 Tg N/yr is consumed by animals in feed, of which about 2 Tg N/yr is emitted to the atmosphere as NH_3 and about 1 Tg N/yr is consumed by humans in meat products (Howarth et al., 2002). Once emitted, the NH_3 can be converted rapidly to ammonium (NH_4^+) aerosol by reactions with acidic species (e.g., HNO_3 [nitric acid], H_2SO_4 [sulfuric acid], NH_4HSO_4 [ammonium bisulfate]). Gaseous NH_3 is removed primarily by dry deposition; aerosol NH_4^+ is primarily removed by wet deposition. The residence time of NH_3 and NH_4^+ in the atmosphere is on the order of days, and they can be transported hundreds of kilometers. As an aerosol, NH_4^+ contributes directly to PM2.5 and, once removed, contributes to ecosystem fertilization, acidification, and eutrophication. Once NH_3 (or NO) is emitted to the atmosphere, each nitrogen atom can participate in a sequence of effects, known as the nitrogen cascade, in which a molecule of NH_3 can, in sequence, impact atmospheric visibility, soil acidity, forest productivity, stream acidity, and coastal productivity (Galloway and Cowling, 2002). Excess deposition of reactive nitrogen (either NH_3 - NH_4^+ or nitrate) can reduce the biodiversity of terrestrial ecosystems (National Research Council, 1997).

Nitric Oxide

Although nitric oxide was not specifically addressed by EPA (2001a), the committee believes it should be included in this report because NO is a precursor to photochemical smog and ozone (O_3), and is oxidized in the atmosphere to nitrate, which along with NH_3 contributes to both fine PM and excess nitrogen deposition. The environmental consequences of nitrate deposition are similar to those of NH_3. NO and nitrogen dioxide (NO_2) are rapidly interconverted in the atmosphere and are referred to jointly as NO_x. A small fraction of NH_4^+ and other reduced nitrogen compounds from animal manure is converted to NO by microbial action in soils. Under the new EPA regulation for ozone (0.08 part per million (ppm) 8-hour average), more rural areas will likely violate the standard, and NO emissions from agricultural soils will become more important. Key variables include land use, the amount of NH_4^+ and nitrate being applied to soils, and the emission rate.

Oxides of nitrogen are the key precursors to tropospheric O_3 (part of photochemical smog). NO_x can be incorporated into organic compounds such as peroxyacetyl nitrate (PAN) or further oxidized to nitric acid. The sum of all oxidized nitrogen species (except N_2O) in the atmosphere is often referred to as NO_y. The residence time of NO_y is on the order of 1 day, unless it is lofted into the free troposphere where the lifetime is longer and environmental effects are more far reaching. Gas-phase HNO_3 can be converted to nitrate aerosol, a contributor to PM2.5, and reduced visibility. Nitric acid and particulate nitrate are removed from the atmosphere by wet and dry deposition with the ecological consequences outlined earlier.

Anthropogenic activities account for most of the NO released into the atmosphere, with combustion of fossil fuels representing the largest source (van Aardenne et al., 2001). Nitrification in aerobic soils appears to be the dominant pathway for agricultural NO release, with only minor emissions directly from livestock or manure. The contribution of soil emissions to the global oxidized nitrogen budget is on the order of 10 percent. Where corn is grown extensively, the contribution is much greater, especially in summer; Williams et al. (1992) estimated that contributions from soils amount to about 26 percent of the emissions from industrial and commercial processes in Illinois and may dominate emissions in Iowa, Kansas, Minnesota, Nebraska, and South Dakota. The fraction of fertilizer nitrogen released as NO_x depends on the mass and form of nitrogen (reduced or oxidized) applied to soils, the vegetative cover, temperature, soil moisture, and agricultural practices such as tillage.

Hydrogen Sulfide

Hydrogen sulfide is produced in anaerobic environments from the decomposition of sulfur-containing organic matter and the reduction of sulfate. It is emitted during manure decomposition and by the reduction of sulfate in feeds and water.

On a global basis, 0.4 - 5.6 Tg S/yr of reduced sulfur gases (mostly H_2S and $(CH_3)_2S$ [dimethyl sulfide] are emitted from land biota and soils (Penner et al., 2001). Most H_2S in the atmosphere is oxidized to sulfur dioxide (SO_2), which is then either dry deposited or oxidized to aerosol sulfate and removed from the atmosphere primarily by wet deposition. The residence time of H_2S and its reaction products is on the order of days. While the terrestrial emissions of H_2S are small compared to SO_2 from fossil fuel combustion (90 Tg S/yr), emissions from AFOs may be important on a local and regional basis. Their effects include an impact on occupational health and a contribution to regional sulfate aerosol loading.

H_2S is regulated (differently) in a number of states (Table 1-1). EPA does not currently list it as a hazardous air pollutant. Because toxic effects depend on both concentrations and exposure times, the periods over which measurements are to be averaged are also shown in Table 1-1.

TABLE 1-1. Current Hydrogen Sulfide Standards in Various States

State	Standard (ppb)	Averaging Period
California	8[a]	Not specified
California	30	1 hr
Illinois	10	8 hr
Minnesota	7	3 months
Minnesota	60	1 hr
New York	0.7	1 yr

[a] Termed the chronic reference inhalation standard. Units are parts per billion.
SOURCE: Environmental Health Sciences Research Center (2002)

INTRODUCTION

Nitrous Oxide

Nitrous oxide is emitted to the atmosphere from animal manure via the processes of nitrification and denitrification. Biogenic sources dominate global N_2O emissions, and of the total 18 Tg N/yr, anthropogenic processes account for about 8.1 Tg N/yr. Of these, cattle feedlots are thought to contribute about 2.1 Tg N/yr and agricultural soils receiving manure about 4.2 Tg N/yr (Prather et al., 2001). N_2O is lost from the troposphere primarily by diffusion into the stratosphere, where it is lost to photolysis and other processes. Once emitted, N_2O is globally distributed because of its long residence time (~100 years) and contributes to both tropospheric warming and stratospheric ozone depletion.

Methane

Methane is produced by microbial degradation of organic matter under anaerobic conditions. Biogenic sources dominate the global CH_4 budget with roughly 60 percent of the total being anthropogenic. Of the global source strength, 600 Tg CH_4/yr, ruminants (domesticated and wild) contribute about 90 Tg CH_4/yr, landfills about 40 Tg CH_4/yr, and rice cultivation about 60 Tg CH_4/yr (Prather et al., 2001). A small portion of U.S. CH_4 emissions come from crop residue burning, wildfires, and wetland rice cultivation. The role of AFOs, especially anaerobic manure lagoons, remains uncertain. Because of the long residence time (~8.4 years) CH_4 becomes distributed globally. Its primary loss mechanism in the atmosphere is conversion to CO. Methane is a greenhouse gas and contributes to global warming (National Research Council, 1992).

The primary source of CH_4 in livestock production is ruminant animals. Globally, domesticated ruminants produce about 80 Tg annually, accounting for about 22 percent of CH_4 emissions from human-related activities (Gibbs et al. 1989). Livestock ruminants (sheep, goats, camel, cattle, and buffalo) have a unique, four-chambered stomach. In one chamber called the rumen, bacteria break down grasses and other feedstuff to generate methane as one of several by-products. Its production rate is affected by several factors (quantity and quality of feed, animal body weight, age, and amount of exercise) and varies among animal species and among individuals of the same species (Leng, 1993).

An adult cow produces between 80 and 120 kg of CH_4 annually. In the United States, cattle emit about 6 Tg CH_4/yr, equivalent to about 4.5 Tg C/yr. Lerner et al. (1988) estimated that of the annual global production of 400 to 600 Tg of CH_4, enteric fermentation in domestic animals contributes approximately 65 to 85 Tg. Methane emissions from agricultural activities in the United States in 1999 were estimated at 9.1 Tg, 32 percent of total U.S. anthropogenic CH_4. Ninety-five percent of CH_4 emissions from agricultural activities came from livestock production. About 65 percent of these emissions could be traced to enteric fermentation in ruminant animals, with the remainder attributable to anaerobic decomposition of livestock manure (DOE, 2000). The most important factor affecting the amount produced by manure is how it is managed, because certain types of storage and treatment systems promote an oxygen-depleted environment. Metabolic processes of methanogens lead to CH_4 production at all stages of manure handling. Liquid systems tend to encourage anaerobic conditions and tend to produce significant quantities of CH_4, while solid waste management approaches may produce little or none. Higher temperatures and moist conditions also promote CH_4 production.

Emissions from agriculture represented about 20 percent of U.S. CH_4 emissions in 1999, with 6 percent from manure. From 1990 to 1999, emissions from this source increased by 8.0 Tg/yr CO_2 (carbon dioxide) equivalent—the largest absolute increase of any of the CH_4 source categories. The bulk of this increase—from swine and dairy cow manure—may be attributed to the shift in composition of the swine and dairy industries towards larger facilities using liquid management systems. Swine manure was estimated to produce 1.1 Tg/yr (CO_2 equivalents), while beef and dairy produce 0.9 Tg/yr (CO_2 equivalents) (EPA, 1999).

Particulate Matter

In the context of this report, particulate matter is grouped into two classes, PM10 and PM2.5. PM10 is commonly defined as airborne particles with aerodynamic diameters less than 10 μm. This definition is not precise, however, and the 10 μm diameter refers to the 50 percent cut diameter in a Federal Reference Method PM10 sampler (Federal Register, 1997), the aerodynamic diameter of a particle collected at 50 percent efficiency. Similarly, PM2.5 refers to the particles that are collected in a Federal Reference Method PM2.5 sampler (Federal Register, 1997) that has a 50 percent cut diameter of 2.5 μm. NAAQS are set for both PM10 and PM2.5 (Table 1-2). AFOs can contribute directly to PM through several mechanisms, including direct emissions from mechanical generation and entrainment of mineral and organic material from the soil and manure or indirect emissions of NO and NH_3 that can be converted to aerosols through reactions in the atmosphere. Ammonium may be a major component of fine particulate matter over much of North America.

The effective aerodynamic equivalent diameter of particulate matter is critical to its health and radiative effects. PM2.5 is targeted because its constituents have the greatest impact on human morbidity and mortality and are most effective in attenuating visible radiation. PM2.5 can reach and be deposited in the smallest airways (alveoli) in the lung, whereas larger particles tend to be deposited in the upper airways of the respiratory tract (National Research Council, 2002). Particles produced by gas-to-particle conversion

TABLE 1-2. National Air Quality Standards for Particulate Matter

Particle Size[a]	Standard ($\mu g/m^3$)	Averaging Period
PM10	50	1 yr
	150	24 hr
PM2.5	15	1 yr
	65	24 hr

[a]PM10 and PM2.5 refer to particulate matter with aerodynamic diameters up to 10 and up to 2.5 µm, respectively.
SOURCE: EPA (2002)

generally fall into the PM2.5 size range. Key variables affecting the emissions of PM10 include the amount of mechanical and animal activity on the dirt or manure surface, the water content of the surface, and the fraction of the surface material in the size range. For PM2.5, key variables affecting the emissions include the net release of precursors such as NO and NH_3.

Volatile Organic Compounds

Volatile organic compounds (VOCs) are organic compounds that vaporize easily at room temperature. They include fatty acids, nitrogen heterocycles, sulfides, amines, alcohols, aliphatic aldehydes, ethers, *p*-cresol, mercaptans, hydrocarbons, and halocarbons. The majority of these compounds participate in atmospheric photochemical reactions, while others play an important role as heat-trapping gases (King, 1995). In 1993, VOC emissions from the San Bernardino Basin from livestock manure were estimated to be 12 tons per day (South Coast Air Quality Management District, 1993). Total emissions of VOCs from all sources in the United States were 30.4 Tg/yr in 1970 and 22.3 Tg/yr in 1995 (EPA, 1995a).

Emission of VOCs from AFOs may cause significant economic and environmental problems. The major constituents that have been qualitatively identified include organic sulfides, disulfides, C_4 to C_7 aldehydes, trimethylamine, C_4 amines, quinoline, dimethylpyrazine, and C_3 to C_6 organic acids in addition to lesser amounts of C_4 to C_7 alcohols, ketones, aliphatic hydrocarbons, and aromatic compounds. Some may irritate the skin, eye, nose, and throat on contact and the mucous membranes if inhaled. VOCs can also be precursors to O_3 and PM2.5. VOCs that cause odors can stimulate sensory nerves to cause neurochemical changes that might influence health by compromising the immune system. Odors associated with VOCs can also trigger memories linked to unpleasant experiences, causing cognitive and emotional effects such as stress. At high levels of exposure, some VOCs are carcinogenic or can cause central nervous system disorders such as drowsiness and stupor.

However, the effects of air emissions from AFOs on public health are not fully understood or well studied. Greater mood disturbance (Schiffman et al., 1995) and increased rates of headaches, runny nose, sore throat, excessive coughing, diarrhea, and burning eyes have been reported by persons living near swine operations in North Carolina (Wing and Wolf, 2000). Thu et al. (1997) observed similarities between the pattern of symptoms among community residents living near large swine operations and those experienced by workers. Caution must be exercised in interpreting the studies because environmental exposure data were not reported.

Odor

Odor is complex both because of the large number of compounds that contribute to it (including H_2S, NH_3, and VOCs), and because it involves a subjective human response. Schiffman et al. (2001) identified 331 odor-causing compounds in swine manure. Though research is under way to relate olfactory response to individual odorous gases, odor measurement using human panels appears to be the method of choice now and for some time to come. Since odor can be caused by hundreds of compounds and is subjective in human response, estimates of national or global odor inventories are meaningless. Odor is also a common source of complaints from people living near AFOs and it is for local impacts that odor has to be quantified.

However, there is some confusion in the literature over how to measure odor intensity. Some define an odor unit (OU) as the mass of a mixture of odorants in 1 m^3 of air at the odor detection threshold (ODT)—the concentration of the mixture that can be detected by 50 percent of a panel. Others define OU as the factor by which an air sample must be diluted until the odor reaches the ODT.

DISTRIBUTION OF EMITTED POLLUTANTS

Temporal Scale

An atmospheric substance can be characterized by its lifetime (also called residence time) in the atmosphere—defined as the time required to reduce its concentration to $1/e$ (e is the base of the system of natural logarithms and has a numerical value of about 2.72; $1/e$ is approximately 0.37) of the initial concentration, with all sources eliminated. The species of interest here span a wide range of lifetimes. Soluble species have lifetimes equivalent to that of water in the atmosphere, about 10 days, depending on precipitation. Reactive species such as NO_x and H_2S have lifetimes on the order of days or less before they are converted to other more water soluble species such as nitric and sulfuric acids. The lifetimes of VOCs are usually controlled by rates of hydroxyl radical (OH) attack, and range from hours to months. The exception is CH_4, with a lifetime of about 8.4 years. N_2O is removed by ultraviolet (UV) photolysis and attack by $O(^1D)$ (an electronically excited oxygen atom generated by O_3 photolysis at wavelengths less than 320 nm) in the stratosphere, and it has a lifetime of about 100 years. N_2O is essentially inert in the troposphere.

Lifetimes vary with location and time. In the planetary boundary layer (PBL)—that part of the atmosphere interacting directly with the surface of the earth and extending to about 2 km—lifetimes tend to be short; below a temperature inversion, dry deposition can rapidly remove reactive species like NH_3. Table 1-3 summarizes typical lifetimes in the PBL for species of interest in this report.

Above the PBL, in the free troposphere where wind speeds are higher, temperatures lower, and precipitation less frequent, the lifetime and range of a pollutant may be much greater. Convection transports short-lived chemicals from the PBL to the free troposphere, where they are diluted by turbulent mixing and diffusion. For key atmospheric species involved in nonlinear processes, such as NO and cloud condensation nuclei (CCN), convection can transform local air pollution problems into regional or global atmospheric chemistry problems.

TABLE 1-3. Typical Lifetimes in the Planetary Boundary Layer for Pollutants Emitted from Animal Feeding Operations

Species	Lifetime
NH_3	~1-10 day
NO_x	~1 day
H_2S	~1 day
N_2O	100 yr
CH_4	8.4 yr
PM	1-10 days, depending on particle size and composition
VOCs	hours to months, depending on compound
Odor[a]	

[a] Odor, which is based on olfactory response to a mixture of compounds, decreases with time in response to dispersion (dilution), deposition, and chemical reactions.

Spatial Scale

Atmospheric concentrations depend on emission or formation rates, loss rates, and mixing, which in turn depend on atmospheric conditions and local geography. Local pollution episodes generally occur with low horizontal wind speeds, as is often the case when a high-pressure ridge dominates the synoptic-scale weather. Inhibited vertical mixing also contributes to high surface concentrations. A strong temperature inversion (temperature increasing rapidly with elevation) at low altitude leads to a shallow PBL and prevents transport of pollutants to the free troposphere. Local concentrations are generally highest when ground-level inversions are strongest. A variety of processes, including subsidence, radiation, and advection, can cause inversions. A detailed discussion is beyond the scope of this report. Local orographic conditions, such as lying in a valley, can exacerbate inversions. Long-lived chemicals such as CH_4 and N_2O can have large-scale (global) effects, but their local concentrations are not usually a problem.

The complexities of the various kinds of air emissions and the temporal and spatial scales of their distribution make their direct measurement at the individual AFO level impractical other than in a research setting. Relatively straightforward methods for measuring emission rates by measuring airflow rates and the concentrations of emitted substances are often not available. Flow rates and pollutant concentrations may be available for some types of confined animal housing but usually not for emissions from soils.

2

Determining Emission Factors

INTRODUCTION

The U.S. Environmental Protection Agency (EPA) has asked the committee to address a number of specific questions (see Executive Summary) relative to characterizing emissions from animal feeding operations (AFOs). The committee has addressed these questions based on the following assumptions developed in earlier sections of this report: (1) emissions estimates are needed at the individual AFO level (Finding 2); (2) it is not practical to measure emissions at all individual AFOs (Finding 3); (3) therefore a modeling approach to predict emissions at the individual AFO level has to be considered; and (4) it is necessary to establish the set of independent variables that are required to characterize AFO emissions at the individual AFO level (Finding 4).

Most local, state, and federal agencies rely on emission factors to develop emission inventories for various substances released to the atmosphere. As defined by the Emission Factor and Inventory Group in the EPA Office of Air Quality Planning and Standards, an emission factor is (EPA, 1995b):

> A representative value that attempts to relate the quantity of a pollutant released to the atmosphere with an activity associated with the release of the pollutant.

Emission factors are generally expressed as mass per unit of activity related to generating the emission per unit time or instance of occurrence. EPA (2001a) proposed defining emission factor as the mass of the substance emitted per animal unit (AU) per year. EPA and the USDA have different definitions of AU (see Appendix B). Throughout this report, the EPA definition is used.

Emission factors are usually derived from calculations based on measured data. Actual measurements of concentrations and flow rates yield a value for an emission rate, the mass of a substance emitted per unit time (e.g., kilograms of ammonia [NH_3] per year). Sometimes it is more appropriate to measure the flux of an emitted substance, the mass emitted per unit area of the source per unit time (e.g., kilograms of NH_3 per hectare-year). An emission rate can be estimated from flux measurements by integrating emissions over the whole area of the emitting source. Emission rates for an AFO can be estimated from emission factors through the simple expression in Equation 2-1:

$$ER = AU \times EF, \qquad \text{(Eq. 2-1)}$$

where ER is the emission rate, AU is the number of animal units associated with the source, and EF is the emission factor in units of mass per AU per unit time. Equation 2-1 illustrates that the uncertainty contained in the numerical values selected for AU and EF are also present in the derived values for ER.

The main goal of the approach outlined by EPA (2001a) is to develop a method for estimating emissions at the individual AFO level that reflects the different kinds of animal production units commonly used in commercial-scale animal production facilities. Specifically, the approach attempts to subdivide the populations of AFOs according to the different production or manure management systems that are commonly used and to develop emission factors for model farms characterized by the processing steps. Assignment of emission factors to each of the individual processing steps within a model farm leads to an estimate of the annual mass of emissions. An estimate of the emissions from an individual AFO can then be made by associating it with the proper model farm, accounting for the AUs housed there, and adding the contributions from the processing steps (housing, manure storage, and land application).

The central assumption of this approach is that the individual processing steps within each identified manure management system are the principal factors that influence emissions. In other words, although there is inherent variability in emissions within each processing step that constitutes a manure management system, the act of subdividing the AFO population into model farms succeeds in decreasing this inherent variability to the point that single emission factors for individual processing steps, when combined, can adequately describe emissions from a model farm and thus from individual AFOs that are assigned to a given model farm category. It is further implied in this approach that the dominant factor controlling the magnitude of the calculated emissions is the number of AUs housed and not other unaccounted-for or unknown factors. This also explains the emphasis on finding the correct emission factors for the individual processing steps since there is an implied supposition that such unique values must exist (EPA, 2001a).

The data quality objectives (defined as the quality of data that will be necessary to solve a problem or provide useful information; Kateman and Pijers, 1981) required to meet the needs of the EPA Office of Air and Radiation are not specified by EPA (2001a). Whatever method is eventually selected to estimate emissions from individual AFOs, the derived estimate will contain some degree of uncertainty. Here the committee emphasizes the data quality that can be assigned to measurements of emissions, and to subsequently derived emission rates and emission factors. This discussion is placed in the context of the five specific questions from the EPA.

SCIENTIFIC CRITERIA

What are the scientific criteria needed to ensure that reasonable and appropriate estimates of emissions are obtained? In this report, "reasonable and appropriate estimates of emissions" is taken to mean emission estimates with acceptable estimates of uncertainty. For emission rates from AFOs—as with all numerical measurements and numerical calculations based on them—uncertainty can be described in terms of accuracy and precision (Taylor, 1987).

Accuracy

In this report "accuracy" is taken to mean the measure of systematic bias in the average of a set of measurements or estimates, and "precision" is taken as the measure of overall reproducibility. Systematic bias can arise from the measurement technology selected to characterize concentrations or from the selection of AFOs that are not representative of the larger population. Typically, concerns about accuracy are limited to the calibration of the analytical instrumentation used. While accurate calibration is an important component of the measurement process, it does not address the possibility that the analytical instrumentation selected may be ill-suited for the task or that bias may be introduced by the experimental design. Possible sources of systematic bias that should be considered include a predominance of daytime sampling when emissions are often higher; ignoring times during the year when buildings are empty; sampling locations that are not representative of exhaust air composition; odor panel sensitivities; and lack of adequate background sampling, especially at larger facilities with multiple housing units in close proximity. The representativeness of the emission factors reported in the scientific literature and used by EPA (2001a) is a major concern since the EPA's Office of Air and

Radiation has no criteria for how to select the AFOs whose measurements are to be used (e.g., whether the AFO was being operated optimally or not), nor have AFOs been chosen at random. Management of an AFO can have a significant impact on its emissions. AFOs at which individual emission measurements have been made have been selected largely based on access (finding operators willing to allow access to their facilities) and the physical characteristics of the sites (as required by criteria associated with the emission measurement technique selected). Thus, calculating a mean emission factor from screened published data by no means guarantees that the calculated value is representative of the AFO population.

Because there are no universally accepted analysis methods, the presence of systematic bias in emission measurements is best evaluated via intercomparison studies in which emissions are determined by two or more separate analytical techniques with differing overall experimental designs. An assessment of accuracy can also be made through the use of elemental (nitrogen [N], carbon [C], or sulfur [S]) mass balances. Nutrient excretion factors (see Appendix B) offer an independent means to set upper limits on possible emission rates. Reported emission rates in excess of nutrient excretion rates should be viewed with suspicion; they may indicate measurement conditions atypical of normal operation, or a fatal flaw in the overall experimental design or instrumentation used in the study.

Precision

Assigning an estimate of precision to measurements of concentrations emitted from different components in a manure management system is not a simple task. One method is to make paired observations with similar instrumentation over the same space and time (Cochran, 1977). The variance is then obtained as follows:

$$\sigma^2 = \frac{1}{n}\sum_{i=1}^{i=n}\frac{(A_i - B_i)^2}{2}, \qquad \text{(Eq. 2-2)}$$

where A_i and B_i represent the ith pair of observations and n represents the number of pairs (Cochran, 1977). This approach often requires duplication of equipment that may not be possible. Spatial variations in emissions may also become important for area sources such as lagoons or cropland receiving manure or lagoon water. Robarge et al. (2002) applied Equation 2-2 (with $n = 90$ paired observations) to estimate precision, expressed as percent coefficient of variation (CV) associated with ambient atmospheric concentrations of gaseous and particulate species measured using annular denuder technology (Purdue,

1992). For ammonia (NH_3) and sulfur dioxide (SO_2), the calculated CV was <10 percent. For nitrous (HONO) and nitric (HNO_3) acids, the CV values were 17.5 and 31 percent, respectively; for particulate ammonium (NH_4^+), sulfate (SO_4^{2-}), and nitrate (NO_3^-), CVs were 13, 18, and 25 percent, respectively.

Determining the precision of emission concentration measurements is also complicated by the fact that such measurements are actually part of a time series with a substantial degree of covariance between measurements. Emissions of gaseous chemical species are highly dependent on microbial decomposition and conversion processes and on physical transport across air-liquid or air-solid interfaces. These processes are in turn dependent on temperature, and variations in temperature are not random but are autocorrelated. The presence of a significant degree of positive autocorrelation in data requires corrections of the standard error of the mean. The variance is underestimated if it is calculated using standard statistical formulas (Code of Federal Regulations, 2001).

The presence of autocorrelation in emissions data also suggests reconsideration of the sampling frequency in order to characterize emissions. Limiting sampling to one or several short series of sequential measurements (as is often done to reduce cost) may in fact be an inefficient and possibly ineffective way to determine actual diurnal or seasonal variations of emissions with time.

Assigning an estimate of precision to an emission factor for an individual AFO is more challenging than assigning it to a set of concentration and airflow measurements. The relative uncertainty associated with emission factors from individual AFOs can be obtained by remembering that emission factors are an estimate of emissions of particulate matter (PM) or a chemical species from a source. According to Equation 2-1, multiplying an emission factor by the AU, yields an emission rate. Integration of the emission rate over time (e.g., one year) yields the total mass emission from the source. For AFOs the total mass emission for a gaseous species containing nitrogen, carbon or sulfur must be a percentage of the total amount of that element excreted. If the individual AFO is in a steady state with regard to the excreted elements nitrogen, carbon, and sulfur, then the percent emissions of these elements should be relatively constant when averaged across several years. A certain percentage is retained for periods longer than one year (e.g., sludge accumulated at the bottom of treatment lagoons), but most of the elements excreted are applied to agricultural land for row crops and grasses, with the remainder emitted as gases or lost in leachate.

The percentage of an excreted element lost as air emissions must fall between 0 and 100 percent, and it is highly unlikely to be at either extreme. Adoption of nutrient management plans further decreases the range of potential emission, since a certain percentage of the excreted nutrients will be used to support crop growth. The problem of determining the relative uncertainty associated with emissions from an individual AFO, then reduces to determining the variation in the percentages of nitrogen, carbon, and sulfur lost from year to year. By way of example, if 60 percent of the excreted nitrogen on a swine AFO is assumed to be emitted to the atmosphere as NH_3 (the value of 60 percent is selected for illustration purposes only and is not a value endorsed by the committee to be used to characterize AFOs), a 1 percent CV associated with this number would mean an uncertainty of ±0.6 percent, while a 10 percent CV would mean an uncertainty of ±6 percent. Given the dependence of NH_3 volatilization on ambient air temperature, it is highly unrealistic to expect uncertainties of 1 percent CV; such uncertainties can be approached only in a laboratory environment. Values of CV of 10 percent or greater are probably much more realistic for real AFOs.

Continuing with the example of 60 percent of the excreted nitrogen emitted as NH_3, the range in uncertainty in emissions, and therefore calculated emission factors, associated with a 10 percent CV can be calculated directly based on the amount of nitrogen excreted and the number of animal units housed. For a finisher swine operation housing 10,000 head (4,000 AUs; 2.5 head per AU), the annual amount of nitrogen excreted is 1.37×10^5 kg using a nitrogen excretion factor of 13.7 kg N/yr per head (Doorn et al., 2002). (This nitrogen excretion factor assumes that 70 percent of nitrogen intake is excreted.) If 60 percent of excreted nitrogen is emitted as NH_3, these numbers translate into an emission factor of 20.6 kg N/AU per year. Although the actual variation is not known, for the purpose of this example, a CV of 10 percent will be assigned, yielding a standard deviation of ±2.1 kg N/AU per year. Given a normal distribution in the percentage of excreted nitrogen lost as NH3, 95 percent (approximately two times the standard deviation) of the derived emission factors for this single AFO fall in the range of 16.4 to 24.8 kg N/AU per year. Carrying through the same calculations, and assuming instead that 80 percent of excreted nitrogen is released as ammonia, yields emission factors ranging from 21.9 to 32.9 kg N/AU per year.

As noted above, these calculations are for illustration purposes only to demonstrate how a relatively modest variation in emissions from a single AFO (10 percent CV) translates into a range of potential emission factors. Yearly variations in emissions are to be expected and cannot be ignored. After careful evaluation of ammonia emissions from swine houses by various methods, Doorn et al. (2002) recommended a general emission factor for houses of 3.7 ± 1.0 kg NH3/yr per finished hog, which is a 27 percent CV. Groot Koerkamp et al. (1998) reported CVs ranging from 17 to 49 percent for different livestock and housing systems in England, the Netherlands, Denmark, and Germany, with between-season CVs ranging from 24 to 57 percent. Although the yearly variation in emissions from single AFOs is not well characterized, the assumed value of 10 percent CV used in the above calculations appears quite conservative compared to these measures of precision reported.

Viewing emissions as a percentage of an element excreted offers a means of estimating the relative uncertainty associated with emissions from individual AFOs. The approach will be most successful for those gaseous species (NH_3, CH_4 [methane], or H_2S [hydrogen sulfide]) whose emissions comprise a substantial portion of the element (nitrogen, carbon or sulfur) excreted. For gaseous species whose emissions represent relatively minor fractions of these excreted elements (e.g., volatile organic compounds [VOCs]), the percent emission becomes less certain, but the approach still makes it possible to set an upper limit on emissions, and the use of percent CV values to estimate relative uncertainty still applies. This approach cannot be used for PM, whose emissions are not a direct function of the amount of a given element excreted, nor can it be applied to odors.

In summary, to ensure that reasonable and appropriate estimates of emissions are obtained from AFOs, the measured and derived emission values must have accompanying measures of uncertainty, including accuracy and precision. Accuracy does not depend simply on instrument calibration; representativeness must be considered since AFOs may not be selected at random and there are no standard methods for measuring emissions. All measurements of emissions should be assumed to have systematic bias and should be compared to other measurements or derived data, such as excretion factors and mass balances. Methods to obtain an estimate of precision do exist and should be included in experimental designs. Short-term sequential measurements will undoubtedly be autocorrelated, and deriving estimates of precision by applying normal statistical techniques to such data will underestimate uncertainties. There are methods for deriving estimates of variance from highly autocorrelated data (Code of Federal Regulations, 2001).

PUBLISHED LITERATURE

What are the strengths, weaknesses and gaps of published methods to measure specific emissions and develop emission factors that are published in the scientific literature?

Ammonia

Several well-designed research studies have been published establishing some of the factors that contribute to variations in NH_3 emissions. For example, Groot Koerkamp et al. (1998) reported wide variations in emissions for different species (cattle, sows, and poultry) measured in different European countries, across facilities within a country, and between summer and fall. Amon et al. (1997) demonstrated that emissions increase as animals age.

Differences due to the manure storage system have been demonstrated (Hoeksma et al. 1982). Climate, including temperature and moisture, also affects NH_3 emissions (Hutchinson et al., 1982; Aneja et al., 2000). Zhu et al. (2000) reported diurnal variation in emission measurements. With so many sources of variation in NH_3 emissions, it is unreasonable to apply a factor determined in one system, over a short period of time, to all AFOs within a broad classification.

Although NH_3 emissions have been reported under different conditions, there are few reliable data to estimate total NH_3 emissions from all AFO components for all seasons of the year. Twenty-seven articles were used for NH_3 emission factors by EPA (2001a); of these, only eleven with original measurements were from peer-reviewed sources. Additional data were taken from six progress reports from contract research. Two of these (Kroodsma et al., 1988; North Carolina Department of Environmental and Natural Resources, 1999) were identified as "preliminary," and in one case (Kroodsma et al., 1988), the airflow measurement equipment was not calibrated.

Emission factors for NH_3 were also taken from nine review articles (EPA, 2001a); three of these modeled or interpreted previously reported information with the objective of determining emission factors (Battye et al., 1994; Grelinger, 1998; Grelinger and Page, 1999). Several of the reviews reported factors used in other countries, but not the original research used to develop them. Other reviews summarized data from primary sources that were already considered. Thus, the review articles may not provide new information.

Most measurements and estimates reported did not represent a full life cycle of animal production. As animals grow or change physiological state, their nutrient excretion patterns vary, altering the NH_3 volatilization patterns (Amon et al., 1997). A single measurement over a short period of time will not capture the total emission for the entire life cycle of the animal. In addition, most measurements for manure storage represent only part of the storage period. The emissions from storage vary depending on length of storage, changing input from the animal system, and seasonal effects such as wind, precipitation (Hutchinson et al., 1982), and temperature (Andersson, 1998). Only one article reported measurements over an entire year (Aneja et al., 2000), although the measurements may not have been continuous. In this case, NH_3 emissions were measured from an anaerobic lagoon using dynamic flow-through chambers during four seasons. Summer emissions were 13 times greater than those in winter, and the total for the year was 2.2 kg NH_3-N per animal (mean live weight = 68 kg) per year.

Expressing NH_3 emission factors on a per annum and per AU basis facilitates calculation of total air emissions and accounts for variation due to size of AFOs, but it does not account for some of the largest sources of variation in emissions. Clearly, there is a great deal of variation in reported measurements among AFOs represented by a single model. For example, only two references were provided for beef drylot NH_3 emission factors, but the values reported were 4.4 and 18.8 kg N/yr per animal (See Table 8-11, EPA, 2001a). For swine operations with pit storage, mean values reported in eight studies ranged from 0.03 to 2.0 kg/yr per pig of less than 25-kg body weight (See Table 8-17, EPA 2001a). This higher rate represents 66 percent of the nitrogen estimated to be excreted by feeder pigs per year (See Table 8-10, EPA, 2001a). The actual variation among AFOs represented by a single model cannot be determined without data representing the entire population of AFOs to be modeled. This would require greater replication and geographic diversity. Much of the variation among studies within a single type of model farm can be attributed to different geographic locations or seasons and the different methods and time frames used to measure the emission factors.

The approach in EPA (2001a) was to average all reported values in selected publications—both refereed and non-refereed—giving equal weight to each article. Emission factors reported in some represented a single 24-hour sample, while in others, means of several samples were used. Emission factors from review articles were averaged along with the others. Properly using available data to determine emission factors, if it could be done, would require considering the uniqueness and quality of the data in each study for the intended purpose and weighting it appropriately. The causes of the discrepancies among studies would also have to be investigated.

Adding emissions from housing, manure storage, and field application, or using emission factors determined without considering the interactions of these subsystems, can easily provide faulty estimates of total emissions of NH_3. If emissions from a subsystem are increased, those from other subsystems must be decreased. For example, most of the excreted nitrogen is emitted from housing, much of the most readily available nitrogen will not be transferred to manure storage. If emissions occur in storage, there will be less nitrogen for land application. The current approach ignores these mass balance considerations, and simply adds the emissions using emission factors determined separately for each subsystem.

Dividing the total manure nitrogen that leaves the farm by the total nitrogen excreted can identify some potential overestimation of emission factors. For example, using emission factors in Table 8-21 of EPA (2001a) for swine model farms, the total ammonia nitrogen emissions for 500 AUs in Model S2 can be estimated to be 1.12×10^4 kg/yr. (Three significant digits are carried for numerical accuracy from the original reference and may not be representative of the precision of the data.) The total nitrogen excreted by 500 AUs of growing hogs is 1.27×10^4 kg/yr (EPA, 2001a). Thus, one calculates that 90 percent of estimated manure nitrogen is volatilized to ammonia, leaving only 10 percent to be accumulated in sludge, applied to crops, and released as other forms of nitrogen NO [nitric oxide], N_2O [nitrous oxide], and N_2). Thus, these emission factors suggest that almost all excreted nitrogen is lost as NH_3, which seems unlikely.

Nitric Oxide

Although nitric oxide was not specifically mentioned in the request from the EPA, the committee believes that it should be included in this report because of its close relationship to ammonia. An appreciable fraction of manure nitrogen is converted to NO by microbial action in soils and released into the atmosphere. NO participates in a number of processes important to human health and the environment. The rate of emission has been widely studied but is highly variable, and emissions estimates are uncertain.

Attempts to quantify emissions of NO_x from fertilized fields show great variability. Emissions can be estimated from the fraction of the applied fertilizer nitrogen emitted as NO_x, but the flux varies strongly with land use and temperature. Vegetation cover greatly decreases NO_x emissions (Civerolo and Dickerson, 1998); undisturbed areas such as grasslands tend to have low emission rates, while croplands can have high rates. The release rate increases rapidly with soil temperature—emissions at 30°C are roughly twice emissions at 20°C.

The fraction of applied nitrogen lost as NO emissions depends on the form of fertilizer. For example, Slemr and Seiler (1984) showed a range from 0.1 percent for $NaNO_3$ (sodium nitrate) to 5.4 percent for urea. Paul and Beauchamp (1993) measured 0.026 to 0.85 percent loss in the first 6 days from manure nitrogen. Estimated globally averaged fractional applied nitrogen loss as NO varies from 0.3 percent (Skiba et al., 1997) to 2.5 percent (Yienger and Levy, 1995). For the United States, where 5 Tg of manure nitrogen is produced annually, NO_x emissions directly from manure applied to soil are roughly 1 percent or 0.05 Tg/yr, neglecting emissions from crops used as animal feed. Williams et al. (1992) developed a simplified model of emissions based on fertilizer application and soil temperature. They estimated that soils accounted for a total of 0.3 Tg or 6 percent of all US NO_x emissions for 1980.

Natural variability of emissions dominates the uncertainty in the estimates. In order of increasing importance, errors in land use data are about 10-20 percent, and experimental uncertainty in direct NO flux measurements is estimated at about ±30 percent. The contribution of soil temperature to uncertainty in emissions estimates stems from uncertainty in inferring soil temperature from air temperature and from variability in soil moisture. Williams et al. (1992) show that their algorithm can reproduce the observations to within 50 percent. A review of existing literature indicates that agricultural practices (such as the fraction of manure applied as fertilizer, application rates used, and tillage) introduce variability in NO emissions of about a factor of two.

Variability of biomes to which manure is applied (such as short grass versus tallgrass prairie) accounts for an additional factor of three (Williams et al., 1992; Yienger and Levy, 1995; Davidson and Klingerlee, 1997). Future research may have to focus on determining the variability of emissions, measured as a fraction of the applied manure nitrogen, with agricultural practices, type of vegetative cover, and meteorological conditions.

Hydrogen Sulfide

Most of the studies on hydrogen sulfide emissions from livestock facilities were conducted recently and included current animal housing and manure management practices. Several recent publications from Purdue University document H_2S emissions from mechanically ventilated swine buildings (Ni et al., 2002a, 2002b, 2002c, 2002d). A pulsed fluorescence SO_2 analyzer with an H_2S converter was used to measure H_2S concentrations in the air, and a high-frequency (16 or 24 sampling cycles each day) measurement protocol was used for continuous monitoring. In one of the studies reported, H_2S emission from two 1,000-head finishing swine buildings with under-floor manure pits in Illinois was monitored continuously for a six-month period from March to September 1997. Mean H_2S emission was determined to be 0.59 kg per day, or 6.3 g per day per 500-kg animal weight. Based on emission data analysis and field observation, researchers noticed that different gases had different gas release mechanisms. Release of H_2S from the stored manure, similar to carbon dioxide and sulfur dioxide, was through both convective mass transfer and bubble release mechanisms. In comparison, the emission of NH_3 was controlled mainly by convective mass transfer. Bubble release is an especially important mechanism controlling H_2S emission from stirred manure. The differences in release mechanisms for different gases are caused mainly by differences in solubility and gas production rates in the manure. Some measurements from swine buildings were also conducted in Minnesota (Jacobson, 1999; Wood et al., 2001).

Very few data are available on H_2S emission from other types of livestock facilities, such as dairy, cattle, and poultry. Using emission data from swine operations to estimate emission factors for other species such as dairy and poultry is not scientifically sound. Outside manure storage, such as storage in tanks or anaerobic lagoons, can be important sources of H_2S emissions. Emission data for such sources are lacking in the literature.

EPA (2001a) stated that H_2S emissions from solid manure systems—such as beef and veal feedlots, manure stockpiles, and broiler and turkey buildings—were insignificant, based on the assumption that these systems are mostly aerobic. Such an assumption is not valid because it is not based on scientific information. Published data indicate that a significant amount of H_2S

is emitted from the composting of poultry manure when the forced aeration rate is low (Schmidt, 2000). It is very likely that H_2S is emitted from other solid manure sources as well. H_2S is produced biologically whenever there are sulfur compounds, anaerobic conditions, and sufficient moisture. Wet conditions occur in animal feedlots and uncovered solid manure piles during precipitation or in rainy seasons. Scientific studies should be conducted to provide emission data.

Nitrous Oxide

Nitrous oxide is both a greenhouse gas and the main source of stratospheric NO_x, the principal sink for stratospheric ozone; predominately biological processes (nitrification and denitrification) produce N_2O in soils; fertilization increases emissions. Although EPA (2001a) states that "emission factors for N_2O were not found in the literature," a large body of research exists on N_2O emissions from livestock, manure, and soils. Time constraints prevent a thorough review of the literature, but this section condenses the main points of a few recent papers and attempts to summarize the state of the science.

N_2O emissions were reviewed for the Intergovernmental Panel on Climate Change (Intergovernmental Panel On Climate Change, 2001; see also Mosier et al., 1998) with the objective of balancing the global atmospheric N_2O budget and predicting future concentrations. Although substantial uncertainties exist regarding the source strength for N_2O, agricultural activities and animal production are the primary anthropogenic sources. According to the Intergovernmental Panel on Climate Change (2001) these biological sources can be broken down into direct soil emissions, manure management systems, and indirect emissions. These three sources are about equally strong, each contributing about 2.1 Tg N/yr to the atmospheric N_2O burden. Total anthropogenic sources are estimated to be 8.1 Tg N/yr, and natural sources about 9.9 Tg N/yr, for a total of 18 Tg N/yr (Prather et al., 2001).

Soils

The Intergovernmental Panel on Climate Change estimated soil N_2O emissions as a fraction of applied nitrogen. They assumed that 1.25 percent of all fertilizer nitrogen is released from soils as N_2O, with a range of 0.25 to 2.25 percent. Estimating direct soil N_2O emissions is subject to the same uncertainties as NO emissions. The fraction of applied nitrogen emitted as N_2O varies with land use, chemical composition of the fertilizer, soil moisture, temperature, and organic content of the soil. Of the global value of 2.1 Tg N/yr emitted directly from soils, Mosier et al. (1998), using the Intergovernmental Panel on Climate Change method, estimates that manure fertilizer contributes

0.63 Tg/yr. Using the Intergovernmental Panel on Climate Change method, 5 Tg/yr of manure nitrogen in the United States would yield 0.06 Tg N/yr as N_2O. Li et al. (1996) employed a model that accounts for soil properties and farming practices and concluded that the Intergovernmental Panel on Climate Change method underestimates emissions. They put annual N_2O emissions from all crop- and pastureland (including emissions from manure and biosolids applied as fertilizer) in the United States in the range of 0.9 to 1.1 Tg N/yr, although this number includes what Mosier et al. (1998) refers to as "indirect" sources.

Nitrification is primarily responsible for NO production, but both nitrification and denitrification lead to N_2O release from soils, and both aerobic and anaerobic soils emit N_2O. The following studies show some of the variability in estimates of the efficiency of conversion of manure nitrogen to N_2O emission. Paul and Beauchamp (1993) measured 0.025 to 0.85 percent of manure nitrogen applied to soil in the lab lost as N_2O, but Wagner-Riddle et al. (1997) found 3.8 to 4.9 percent from a fallow field. Petersen (1999) observed 0.14 to 0.64 percent emission from a barley field. Lessard et al. (1996) measured 1 percent emission of manure nitrogen applied to corn in Canada. Yamulki et al. (1998) measured emissions from grassland in England and found 0.53 percent of fecal nitrogen and 1.0 percent of urine nitrogen lost as N_2O over the first 100 days. Whalen et al. (2000) applied swine lagoon effluent to a spray field in North Carolina and observed 1.4 percent emission of applied nitrogen as N_2O. Flessa et al. (1995) applied a mixture of urea and NH_4NO_3 to a sunflower field in southern Germany and measured an N_2O emission of >1.8 percent of the nitrogen applied. Long-term manure application (possibly linked to increased organic content of soils) appears to increase N_2O production. Rochette et al. (2000) determined that after 19 years of manure application, 1.65 percent of applied nitrogen was converted to N_2O. Chang et al. (1998) followed the same soil for 21 years of manure application and found 2-4 percent of manure nitrogen converted to N_2O. Flessa et al. (1996) determined a total emission of N_2O from cattle droppings on a pasture equivalent to 3.2 percent of the nitrogen excreted. Clayton et al. (1994) showed that grassland used for cattle grazing could convert a larger portion of fertilizer ammonium nitrate (NH_4NO_3) nitrogen to N_2O (5.1 percent versus 1.7 percent for ungrazed grassland). Williams et al. (1999) applied cow urine to pasture soil in the lab and observed a 7 percent partition of the nitrogen to N_2O.

Manure Management

Several recent studies indicate that N_2O emissions from manure can be large (Jarvis and Pain, 1994; Bouwman, 1996; Mosier et al., 1996; Intergovernmental Panel on Climate, 2001). For example, Jungbluth et al. (2001) measured 1.6 g N_2O/d per 500 kg of livestock emitted directly from dairy cattle; Amon et al, (2001) measured 0.62 g N_2O/d per 500 kg of livestock. Groenestein and VanFaassen (1996) found 4.8 to 7.2 g N/d per pig as N_2O.

The Intergovernmental Panel on Climate Change (2001) estimates N_2O emissions from animal production (including grazing animals) as approximately 2.1 Tg N/yr. These estimates are based on an assumed average fraction of manure nitrogen converted to N_2O and are subject to variability due to temperature, moisture content, and other environmental factors in a manner similar to soil emissions. Berges and Crutzen (1996) estimated the rate of N_2O emissions by measuring the ratio of N_2O to NH_3. They determined that 40 Tg N/yr of cattle and swine manure in housing and storage systems generates 0.2-2.5 Tg N/yr as N_2O; they did not account for additional emissions outside the housing and storage systems.

Indirect Emissions

Formation of N_2O results indirectly from the release of NH_3 to the atmosphere, and its subsequent deposition as NH_3-NH_4^+ or nitrate, or from their leaching and runoff (Intergovernmental Panel on Climate Change, 2001). Human waste in sewage systems is another indirect path to atmospheric N_2O. On a global scale, leaching and runoff give an estimated 1.4 Tg N/yr; atmospheric deposition, 0.36 Tg N/yr; and human sewage, about 0.2 Tg N/yr—for a total of about 2 Tg N/yr. Dentener and Crutzen (1994) pointed out that atmospheric reactions involving NH_3 and NO_2 could lead to production of N_2O; however the strength of this source is unknown.

Summary

The uncertainty in emissions of N_2O from AFOs is similar to that for NO—roughly a factor of three. While no-till agriculture decreases emissions of most greenhouse gases (Civerolo and Dickerson, 1998; Robertson et al., 2000) it appears to increase N_2O. The means for decreasing emissions do exist. Smith et al. (1997) suggested that substantial reductions in N_2O could be achieved through matching fertilizer type to environmental conditions and by using controlled-release fertilizers and nitrification inhibitors. Timing and placement of fertilizer and controlling soil conditions could also help decrease N_2O production. The vast body of work on emissions of N_2O from agricultural activities cannot be thoroughly reviewed in the short time frame of this study.

Methane

Four original research articles, an agency report, one doctoral thesis, and one review article are cited in EPA (2001a) in estimating emission factors for CH_4. Much research was overlooked since a number of papers and reports describing CH_4 emission rates can be found in the literature. Fleesa et al. (1995) reported CH_4 fluxes of 348 to 395 g per hectare (ha) per year in fields fertilized with manure. A value of 1 kg/m^2 per year CH_4 (carbon equivalents) has been reported for an uncovered dairy yard (Ellis et al., 2001). Amon et al. (2001) concluded that methane emissions were higher for anaerobically treated dairy manure than for composted manure.

EPA (2001a) estimates the CH_4 production potential of manure as the maximum quantity of CH_4 that can be produced per kilogram of volatile solids in the manure. However, a considerable amount of CH_4 is lost during eructation (belching), which this estimate does not take into account.

In estimating the CH_4 emission factor for the model farm, EPA (2001a) did not take several factors into consideration, such as the difficulty associated with measuring emissions without having a negative impact on animals. New methods have been designed to measure CH_4 emissions under pasture conditions with minimal disturbance of the animals (Leuning et al., 1999). There are some limitations to this technique; it does not work well with low wind speeds or rapid changes in wind direction, and requires high-precision gas sensors. Methane production increases while cattle are ruminating (digesting) feedstuffs—both grass and high-energy rations. In one study, lactating beef cattle grazing on grass pasture were observed to have 9.5 percent of the gross energy intake converted to CH_4 (McCaughey et al., 1999). During periods when the cattle are fed a high-grain diet, approximately 3 percent of gross intake energy is converted to CH_4 (Johnson et al., 2000).

Methods for estimating CH_4 emissions from other sources—such as rice paddies, wetlands, and tundra in Alaska—have been well studied. However, the models used to extrapolate emissions over these large areas may not apply to AFOs because of the different variables that must be taken into account. This is a knowledge gap that has to be addressed.

Particulate Matter

A limited number of studies have reported emission factors for particulate matter for various confinement systems. One of the most recent reports includes the results of an extensive study that examined PM emissions from various confinement house types, for swine, poultry, and dairy in several countries in Northern Europe (Takai et al., 1998), and a few studies report cattle or dairy drylot emissions in the United States (Parnell et al., 1994; Grelinger,

1998; Hinz and Linke, 1998; USDA, 2000). Some of this work was cited by EPA (2001a). Two PM10 emission factors for cattle were reported for drylot feed yards by Grelinger (1998) and USDA (2000). Another emission factor for poultry broiler house emissions was also included (Grub et al., 1965).

According to the EPA (1995b) AP-42 document, emission factor data are considered to be of good quality when the test methodology is sound, the sources tested are representative, a reasonable number of facilities are tested, and the results are presented in enough detail to permit validation. Whenever possible, it is desirable to obtain data directly from an original report or article, rather than from a compilation or literature summary. Only a very limited number of published papers have been used to estimate PM emission factors for AFOs. Some of the papers utilized do not appear to be of the highest quality or relevance to modern operations. Takai et al. (1998) and Grub et al. (1965) appeared in the peer-reviewed literature, but other work cited was not. Takai et al. (1998) represents one of the most extensive studies conducted on livestock houses to date; it made 231 field measurements of dust concentrations and dust emissions from livestock buildings across Northern Europe. Factors included in their study design were country (England, the Netherlands, Denmark, and Germany); housing (six cattle housing types, five swine housing types, and three poultry housing types); season (summer and winter); and diurnal period (day and night). Each field measurement was for a 12-hour period, and each house was sampled for a 24-hour period, or two 12-hour samples per house. Where possible, measurements were repeated at the same house for both seasons (Wathes et al., 1998).

One reference (Grelinger, 1998) appeared in a specialty conference proceedings (non-peer reviewed), and it is not clear how the emission rates were derived. USDA (2000) summarizes results from other cattle studies. The Grub et al. (1965) study was more than 35 years old and reported emission factors for a poultry confinement configuration (chambers 2.4 m by 3.0 m by 22.1 m high, ventilated at a constant airflow rate) that is not used in current operations.

The sizes of ambient particulate matter varied from study to study, ranging from "respirable" and "inhalable" to total suspended particulates (TSPs). Takai et al. (1998) sampled inhalable dust using European Institute of Occupational Medicine (IOM) dust samplers. The respirable fraction was measured using cyclone dust samplers with a 50 percent cut diameter of 5 micrometers (μm). Grub et al. (1965) measured dust rather than PM10; it is not clear whether the emission factors quoted represented dust or PM10 estimated from the dust. Grelinger (1998) measured TSP and obtained PM10 by multiplying by 0.25. USDA (2000) reported that TSP was measured rather than PM10, according to the AFO project data summary sheets in EPA (2001a). The representativeness of emission factors in the literature is also questionable. For example, the emission factors reported by Takai et al. (1998) were based on data collected for very brief periods, one to two days at each barn. Relevant work was overlooked in the estimation of cattle feedlot PM emissions (e.g., Parnell et al., 1994), or it is not clear from EPA (2001a) whether that work was included in the USDA (2000) publication cited. Auvermann et al. (2001) extensively reviewed the PM emission factors suggested for AFOs (for both feedlots and feed mills) in AP-42 (EPA, 1995b). They pointed out that the PM10 emission factor for cattle feedlots specified in AP-42 was five times as high as the more recent values determined by Parnell et al. (1994). EPA (2001a) did not discuss the AP-42 emission factors.

When more than one study was found that examined PM emissions, the results were not consistent among studies. The two poultry house emission factors differed by an order of magnitude and were simply averaged to characterize PM emissions from poultry houses, even though the Grub et al. (1965) study was of questionable relevance to today's production systems. The two drylot cattle yard PM emission factors differed by a factor of five and were averaged to characterize the PM emissions from drylots.

Relevant work was overlooked by EPA (2001a) for the estimation of cattle feed yard PM emissions. Recent work by Holmen et al. (2001) using Lidar (light detection and ranging) was not included. Parnell et al. (1994) was not cited, but it is not clear whether that work was included in USDA (2000), which was cited. Potential PM emissions from land spraying with treatment lagoon effluent are assumed to be negligible and thus were not considered further by EPA (2001a).

For PM, unlike most other air pollutants, emission factors developed for use in emission inventories and for dispersion modeling can, ideally, be reconciled using receptor modeling techniques. Receptor modeling makes use of the fact that atmospheric PM is composed of many different chemical species and elements. The sources contributing to ambient PM in an airshed also have specific and unique chemical compositions. If there are several sources and if there is no chemical interaction between them that would cause an increase or decrease, then the total PM mass measured at a "receptor" location will be the sum of the contributions from the individual sources. By analyzing the PM for various chemical species and elements, it should then be possible to back-calculate the contributions from various sources in the airshed. A variety of techniques are available for doing this; some (e.g., the chemical mass balance model; Watson et al., 1997) rely on the availability of predetermined source chemical composition libraries and are based on regression to determine the amounts contributed by various sources. Other receptor models are based on multivariate techniques and do not require source "fingerprints" determined *a priori*, but do require large numbers of receptor samples so that statistical methods can be applied. Target transformation factor analysis (Pace, 1985) and positive matrix factorization (Ramadan et al., 2000) are two examples of multivariate techniques that do not require explicit source composition data. Source apportionment may be especially useful for understanding the contributions from AFOs to the ambient PM in an airshed. Both receptor and dispersion modeling are associated with a significant level of uncertainty. The best approach is to use a combination of methods and attempt to reconcile their results.

Volatile Organic Compounds

Emissions of volatile organic compounds from stationary and biogenic sources are significant, but limited data are available in most regions of the world. This situation makes it difficult to determine the impact of VOCs on a global basis. However, the United States (EPA, 1995a) and Europe have accumulated extensive data on the quantities and sources of their VOCs emitted to the atmosphere.

The three references in EPA (2001a) on VOC emission factors, Alexander, 1977; Brock and Madigan, 1988; and Tate, 1995, came from microbiology textbooks. Thus, the basis for determining VOC emission factors was rather weak.

Despite the paucity of data, attempts are being made to shed light on the estimation of emission factors for VOCs. For example, some for pesticides have been determined by the Environmental Monitoring Branch of the Department of Pesticide Regulation in Sacramento, California (California Environmental Protection Agency, 1998, 1999, 2000). The applicability of these efforts to VOC emissions from AFOs is unknown at this time.

Ongoing studies to determine emission rates of VOCs were not included in EPA (2001a). Scientists from Ames, Iowa, have developed techniques to collect and measure VOCs emitted from lagoons and earthen storage systems (Zahn et al., 1997). They found that 27 VOCs were prevalent in most samples, and could be classified as phenols, indoles, alkanes, amines, fatty acids, and sulfur-containing compounds. Emission rates for many of these were determined at several sites, and the data have been transferred to EPA and state air quality specialists.

According to EPA (2001a), estimation of VOC emissions from confinement facilities, manure storage facilities, and manure application sites is difficult because of the lack of a reasonable method for estimating CH_4 production. CH_4 does not provide an appropriate basis for predicting VOC volatilization potential in livestock management systems. Gas transfer velocities for CH_4 and VOCs differ by several hundredfold (MacIntyre et al., 1995). In addition, surface exchange rates for some VOCs are influenced by solution-phase chemical factors that include ionization (pH), hydrogen bonding, and surface slicks (MacIntyre et al., 1995). Physical factors such as temperature, irradiance and wind are also major factors in the emission rates of sparingly soluble VOCs from liquid or semisolid surfaces (MacIntyre et al., 1995; Zahn et al., 1997). The differences in wind and temperature exposures between outdoor and indoor manure management systems can account for between 51 and 93 percent of the observed differences in VOC emissions (MacIntyre et al., 1995). This analysis suggests that exposure factors can account for differences observed in VOC flux rates, VOC air concentrations, and odor intensities. Therefore, the equation used to model the emission factor for VOCs in EPA (2001a) cannot be extrapolated for the majority of livestock operations.

Receptor modeling techniques can provide information on air quality impacts due to VOC emissions from AFOs. For example, Watson et al. (2001) reviewed the application of chemical mass balance techniques for VOC source apportionment. Multivariate methods have also been applied to source apportionment of ambient VOCs (Henry et al., 1995). Receptor modeling techniques to apportion VOCs from AFOs may be limited because many of the expected compounds may be formed in the atmosphere, react there, or have similar emission profiles from many sources.

To understand the contribution of AFO VOCs to ozone formation and gain insight into effective control strategies, measurements of individual compounds are essential. This is a difficult task because of the large number of compounds involved. The most widely used analytical technique involves separation by gas chromatography (GC) followed by detection using a flame-ionization detector (FID) or mass spectrometer (MS). The latter is useful for identification of non-methane hydrocarbons using cryofocusing. VOC detectors that can be used for real-time measurements of typical ambient air are commercially available. New portable devices that use surface acoustic wave technology have been developed for field measurements of VOCs. Their sensitivity is not adequate to measure the low levels that may be harmful to humans. Research to support the development of more sensitive devices is needed.

There is a lack of information on the acute and chronic toxicological effects of VOCs from agricultural operations on children and individuals with compromised health. Recent epidemiological studies (without environmental measurements of VOCs) have shown higher incidences of psychological dysfunction and health-related problems in individuals living near large-scale swine production facilities (Schiffman et al., 1995; Thu et al., 1997). Further studies are needed to better understand the risks associated with human exposure to VOCs from AFOs.

Odor

In a recent review, Sweeten et al. (2001) define odor as the human olfactory response to many discrete odorous gases. Regarding the constituents of animal odors, Eaton (1996) listed 170 unique compounds in swine manure odor while Schiffman et al. (2001) identified 331. Hutchinson et al. (1982) and Peters and Blackwood (1977) identified animal waste as a source of NH_3 and amines. Sulfides, volatile fatty acids, alcohols, aldehydes, mercaptans, esters, and carbonyls were identified as constituents of animal waste by the National Research Council (1979), Miner (1975), Barth et al. (1984), and the American Society of Agricultural Engineers (1999). Peters and Blackwood (1977) list 31 odorants from beef cattle feedlots. Zahn et al. (2001) found that nine VOCs correlated with swine odor. The sources of odors include animal buildings, feedlots, manure handling, manure storage and treatment facilities, and land applications.

Sweeten et al. (2001) also outline various scientific and engineering issues related to odors, including odor sampling and measurement methods. Odors are characterized by intensity or strength, frequency, duration, offensiveness, and character or quality. Odor concentration is used for odor emission measurement. Several methods are available for measuring odor concentrations including sensory methods, measurement of concentration of specific odorous gases (directly or indirectly), and electronic noses.

Human sensory methods are the most commonly used. They involve collecting and presenting odor samples (diluted or undiluted) to panelists under controlled conditions using scentometers (Huey et al., 1960; Barneby-Cheny, 1987; Miner and Stroh, 1976: Sweeten et al. 1977, 1983, 1991), dynamic olfactometers, and absortion media (Miner and Licht, 1981;Williams and Schiffman, 1996; Schiffman and Williams, 1999). Among sensory methods the Dynamic Triangle Forced-Choice Olfactometer (Hobbs et al., 1999; Watts et al., 1994; Ogink et al.1997) appears to be the instrument of choice. Currently, there is an effort among researchers from several universities, including Iowa State University, the University of Minnesota, Purdue University, and Texas A& M University, to standardize the measurement protocol for odor measurement using the olfactometer.

Some odor emission data are available in the literature, particularly for swine operations (e.g., Powers et al., 1999). However, there are discrepancies among the units used in different studies. Standard measurement protocols and consistent units for odor emission rates and factors have to be developed. As shown in a recent review (Sweeten et al., 2001), the data (see Table 2-1) on odor or odorant emission rates, flux rates, and emission factors are lacking for most livestock species (and for different ages and housing) and are needed for the development of science-based abatement technologies. Further research in well-equipped laboratories is needed as a precursor to rational attempts to develop emission factors for odor and odorants.

TABLE 2-1. Odor Emission Rates from Animal Housing as Reported in the Literature

Animal Type	Location	Odor Emission Flux Rate (OU/s-m^2)[a]	Reference
Nursery pigs (deep pit)	Indiana	1.8[a]	Lim et al., 2001
Nursery pigs[b]	Netherlands	6.7	Ogink et al., 1997; Verdoes and Ogink, 1997
Nursery pigs	Minnesota	7.3-47.7	Zhu et al., 1999
Finishing pigs	Minnesota	3.4-11.9	Zhu et al., 1999
Finishing pigs[c]	Netherlands	19.2	Ogink et al., 1997; Verdoes and Ogink, 1997
Finishing pigs[d]	Netherlands	13.7	Ogink et al., 1997; Verdoes and Ogink, 1997
Finishing pigs (daily flush)[e]	Indiana	2.1	Heber et al., 2001
Finishing pigs (pull-plug)[e]	Indiana	3.5	Heber et al., 2001
Finishing pigs (deep pit)	Illinois	5.0	Heber et al., 1998
Farrowing sows	Minnesota	3.2-7.9	Zhu et al., 1999
Farrowing sows	Netherlands	47.7	Ogink et al., 1997; Verdoes and Ogink, 1997
Gestating sows	Minnesota	4.8-21.3	Zhu et al., 1999
Gestating sows	Netherlands	14.8	Ogink et al., 1997; Verdoes and Ogink, 1997
Broilers	Australia	3.1-9.6	Jiang & Sands, 1998
Broilers	Minnesota	0.1-0.3	Zhu et al., 1999
Dairy cattle	Minnesota	0.3-1.8	Zhu et al., 1999

Note: Rates have been converted to units of OU/s-m^2 for comparison purposes.
[a] Net odor emission rate (inlet concentration was subtracted from outlet concentration).
[b] Number of animals calculated from average animal space allowance.
[c] Pigs were fed acid salts.
[d] Multiphase feeding.
[e] Odor units normalized to European Odor Units based on *n*-butanol.
SOURCE: Adapted from Sweeten et al. (2001).

CHARACTERIZING VARIABILITY

How should the variability in emissions be characterized that is due to regional differences, daily and seasonal changes, animal life stage, and different management approaches? Each model farm proposed by EPA (2001a; Appendix D) includes three variable elements: a confinement area, manure management system, and land application method. The manure management system was subdivided into solid separation and manure storage activities. The model farm assumes that emissions depend primarily on the category identified for each individual element. The potential influences of regional differences, hourly, daily and seasonal changes, animal life stages, and different management approaches are not explicitly considered.

Climatic and Geographic Differences

Differences in climate will influence emissions from AFOs because of differences in temperature, rainfall frequency and intensity, wind speed, topography, and soils. EPA (2001a) notes several possible influences of climatic differences by acknowledging the influence of air temperature on gaseous emissions and the effect of rainfall frequency on stocking densities at cattle and dairy feedlots. Climatic differences per se were excluded from the criteria used to select emission factors from the scientific literature; however the van't Hoff-Arrhenius equation was used to adjust CH_4 conversion factors for mean temperature differences (See Chapter 8; EPA, 2001a).

Increases in mean ambient temperature are expected to increase gaseous emission rates from several components of the model farms, including manure storage and land to which manure has been applied. It is unclear how averaging reported emission factors would remove this influence of temperature, especially if the selected emission factors used were mostly determined in climatic region of the country. The same logic applies to estimates of emissions from housing units or land. Depending on one or two published emission factors from one region of the country results in a possible systematic bias because of climatic differences. This bias is still present when emission factors for one species are applied to others by adjusting them to reflect differences in excretion rates, or by assuming that emissions from an anaerobic poultry lagoon are similar to those from an anaerobic swine lagoon (See Chapter 8, EPA, 2001a).

Differences in emissions from AFOs may also arise because of other geographic differences such as availability of land for manure or lagoon effluent disposal, rates of evapo-transpiration, and differences in soil texture and drainage that can impact application rates of lagoon water, or differences in soil microenvironments that affect microbial action and the resulting gaseous emissions. The breed of a given animal species (e.g., selection for cold or heat tolerance) and feed formulations (due to changes in animal maintenance requirements) may also vary in response to geographic and climatic differences.

It is difficult to project how these various sources of uncertainty will combine to influence gaseous emissions and whether these factors will have significant impact on total percentages of nitrogen, carbon or sulfur lost in gaseous species, when averaged over a year's time. Climatic differences do not negate the mass balance flow of elements through AFOs, so that, unless there is a significant change in storage of an element within the manure management system, changes in total emissions (air and water) can come about only because of changes in excretion (resulting from changes in feed formulation or efficiency of animal nutrient utilization). Differences may not be as important for annual emissions of major gaseous species (such as NH_3 and CH_4) as for VOCs and PM.

Hourly, Daily, and Seasonal Changes

Changes in emissions from individual AFOs due to hourly, daily, and seasonal variations are discussed here because measurements to characterize emissions are usually conducted for short periods of time, preferably during different seasons of the year. Failure to account for short-term cycles in an experimental design used to characterize emissions could result in significant systematic error in a derived emission factor, when extrapolated to a one-year time period.

Individual AFOs are essentially a collection of different biological systems each operating with its own hourly, daily, and seasonal cycles. At the scale of the individual animal, there are daily cycles in activity related to eating, defecating, and moving about (the latter being particularly important for generating PM from cattle feedlots). Microbial cycles that produce emissions may be closely tied to animal activity through the amount and frequency of defecation. As an animal grows, the amount and composition of its feed intake change, as does the amount and composition of its manure (National Research Council, 1994, 1998, 2000, 2001). This gives rise to corresponding changes in total microbial activity and emissions. Lactating animals experience changes in productivity throughout their natural cycle, with changes in feed consumed and nutrients excreted (National Research Council 1998, 2000, 2001). Although the capacity within an AFO remains essentially constant, a number of different animals may occupy this space during the year, depending on the production cycle used. Thus, the cycling of animals through an AFO is another source of variation in emissions.

Upsets in daily rhythms of animals must also be considered, because they may result in changes in feed uptake and nutrients excreted for a period of several days. Such upsets may occur due to illness, drastic short-term changes in weather, or breakdowns of farm equipment. Depending on the manure management system being employed, such event-driven processes may not be significant in terms of emissions of NH_3 or CH_4 but may have a major impact on other emitted species such as VOCs and PM. Other event-driven processes that can occur include lagoon turnover, flush cycles for housing units, and manure scraping at feedlots. As noted by EPA (2001a), these events can result in enhanced emissions.

The impact of daily cycles on emissions is not important when averaged over a yearly time scale, provided a sufficient number of observations are made to account for such cycles. However, given the paucity of emissions data deemed valid for the development of emission factors to characterize the model farms, it is not possible to determine to what extent such cycles may have impacted published emission measurements. As noted earlier, averaging published emission factors does not compensate for the presence of systematic bias that may be present as a result of a failure of the experimental design to account adequately for such cycles.

Animal Life Stage

Reference has already been made to differences in feed formulations that occur during the life cycles of most animals produced at AFOs, and the subsequent effects on the amount and composition of fecal matter and urine excreted. In this section, a specific example is provided (Figure 2-1) of changes in the rate of nitrogen excreted for "grow-finish" swine produced at AFOs in the southeastern United States. The data are based on a growth model (Agricultural Research Council, 1981) used by a commercial swine producer to adjust feed formulations. To prevent the disclosure of proprietary information, data have been normalized to 100 percent for the highest rate of nitrogen excretion per day.

As expected, the relative amount of nitrogen excreted daily tends to increase as the pig grows, reflecting changes in the daily total nitrogen consumed. The actual feed formulation is changed four times during the growth cycle of the hog (not twice as assumed by EPA, 2001a) to account for changes in nitrogen required for maintenance and growth. The changes in the relative amount of nitrogen excreted per day with changes in formulation are not simply an artifact of the model but reflect periods of adjustment by the animal to the changes in feed composition. Overall there is a series of curvilinear increases in the amount of nitrogen excreted per day for finishing swine under this model, with nitrogen excretion nearly doubling during the latter half of the animal's growth period. The emphasis in Figure 2-1 is on total nitrogen excreted. Expressed as a percentage of body weight, the nitrogen excreted would actually be decreasing throughout the growth cycle.

Figure 2-1 illustrates that if daily housing emissions of NH_3 are directly related to daily nitrogen excretion and the model is an accurate representation of nitrogen excretion, then there will not be a simple increase in emissions from the

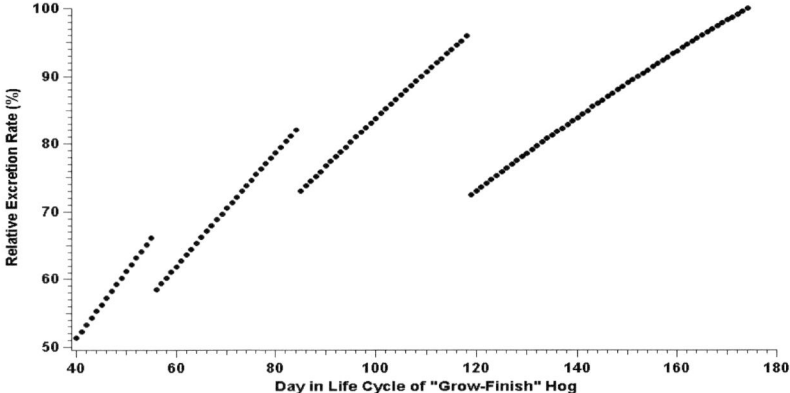

FIGURE 2-1. Relative excretion rate of nitrogen versus day in the life cycle of a grow-finish hog at a commercial swine production facility in the southeastern United States. Animals attain the designation of grow-finish hog at approximately day 40 in their life cycle and are finished at about day 174. Note: Relative excretion rates refer to kilograms of nitrogen per day excreted on day n relative to day 174.

confinement unit with time. Thus, averaging together emission measurements made from several different housing units with different age animals, or from the same housing unit during different times during one growth cycle, may significantly under- or overestimate emissions, depending on the age of the animals when sampled. Actual emissions, however, will also depend on the manure collection practices (flush frequency, pit recharge, pull plug, or pit storage) associated with the confinement unit. A manure collection practice that accumulates manure for relatively long periods of time, such as pit storage, may act to smooth the variations in emissions due to variations in daily excretion of nitrogen. At a minimum the data displayed in Figure 2-1 demonstrate that the same sampling scheme may not be applicable to all swine confinement units and that measurements of emissions may have to be weighted to account for differences in animal age.

Management

Optimal management is vital to the success of individual AFOs for the production of quality animals, and should also result in decreased emissions. Appropriate drainage and manure removal minimizes PM generation from cattle feedlots (Sweeten et al., 1998). Effects of animal health on feeding habits is important to maintain consistent nutrient uptake efficiency and prevent feed spoilage. This attention includes maintenance of proper ventilation for animals in confined housing units, maintenance of drainage systems to remove wastes from housing units on a frequent basis, and regular (perhaps daily) visual inspection of animals and their daily routines. Adherence to nutrient management plans will reduce the potential for of excessive air emissions or surface runoff resulting from overapplication of nutrients to crops. Anaerobic lagoons should not exceed design-loading rates and should be maintained at the proper pH range for waste stabilization.

Assessing the overall quantitative impact of effective management on decreasing emissions is currently not possible due to the paucity of emissions data. However, management practices should not be excluded in assessing emissions from individual AFOs. One way to achieve this goal would be to determine whether managers at AFOs where measurements of emissions are scheduled are in compliance with animal industry guidelines for decreasing emissions, including odors. An illustration of one such program is the America's Clean Water Foundation's (ACWF) On Farm Assessment and Environmental Review (OFAER) project (2002), which reportedly provides livestock producers a confidential, comprehensive, and objective assessment of water quality, odor, and pest risk factors at their operations. (Reference to the OFAER program is for illustration purposes only and should not be construed as an endorsement of this program by the committee or the National Research Council.) The OFAER project currently has the participation of approximately 3,200 AFOs nationwide. Using voluntarily provided emission factors from individual AFOs may produce the database necessary to assess the impact of management on emissions.

In summary, the answer to the question of how the variability in emissions due to regional differences, hourly, daily, and seasonal changes, animal life stage, and different management approaches should be characterized is through consideration of these factors in experimental designs for measuring emissions and deriving emission factors. Average ambient temperatures are the main differences among different regions of the country. Selecting an emission factor based on data from one region (e.g., the southeastern United States) and extrapolating it to other regions or even to other animal types is questionable at best, and must necessarily introduce systematic bias into the derived emission rates for individual AFOs. Because of the importance of temperature effects on microbial activity and gas exchange across different interfaces, accounting for regional differences must include actual measurements of emissions at AFOs across the United States.

Consideration of daily and seasonal changes and animal life stages speaks to the need to consider variations in emissions that occur on the same time scale as most field measurements of emissions at AFOs. Proper characterization of these variations will require experimental designs that encompass the full life cycle of the animals under production and consider whether measured emission rates are nonlinear during the typical animal life cycle. If emissions are in fact nonlinear, then observations of emission rates have to be weighed accordingly when extrapolating to a one-year time frame.

Since AFOs will probably never be chosen at random for field measurements of emissions, selection criteria should be developed for what constitutes an acceptable AFO for field measurements. These criteria should include an evaluation of management and reflect the growing volunteer effort to address water quality and odor and pest issues, for example the ACWF OFAER project.

STATISTICAL UNCERTAINTY

How should the statistical uncertainty in emissions measurements and emissions factors be characterized in the scientific literature? As noted earlier in this chapter, uncertainty can be described in terms of accuracy and precision. Deviations from accuracy (systematic bias) for individual measurement technologies will be addressed in more detail in the final report. This section addresses the broader issue of uncertainty associated with published emissions data and their use in deriving emission factors.

An example of the uncertainty associated with published emission rates from AFOs is illustrated in Table 2-2 adapted from Tables 9 and 10 in a recent review paper (Arogo et al., 2001) summarizing recently published measurements of NH_3 flux (kilograms of NH_3-N per hectare per day) from primary anaerobic swine lagoons. Multiplying the fluxes by the lagoon surface areas gave the daily emission rates for various seasons. The majority of observations listed in Table 2-2 were from "farrow-finish" AFOs, with the remainder from "farrow-wean," "grow-finish," and "breed-wean" facilities. The range in lagoon pH values was 6.8-8.3, but the majority were between 7.4 and 8.2.

The variability in the calculated emission rates in the table is evident in the range of values listed for each combination of measurement method and measurement period, with typical factors of 3 to 7. Seasonal differences in emission rates are also evident, with the ratio of summer to winter rates being as large as 10 or more. Within-lagoon variation in total ammoniacal nitrogen (TAN) is much less, but between lagoons the values vary by factors as high as 10. There is also no obvious association between TAN concentrations in the lagoons and calculated emission rates. The range of rates for individual lagoons is evidence of the uncertainty that must be associated with emission factors derived from published emission rates. Failure to document this uncertainty in tabulated values of emission factors can lead to unrealistic expectations

TABLE 2-2. Calculated Emission Rates of Ammonia from Primary Anaerobic Swine Lagoons as a Function of Measurement Method and Measurement Period

Measurement Method[a]	Period	TAN[b] mg/L	Emission Rate (kg NH_3-N/d)	Reference
Micromet.	Aug-Oct	917-935	29-51	Zahn et al. (2001)
Micromet.	Summer	230-238	11.2-140	Harper et al. (2000)
Micromet.	Winter	239-269	4.6-6.7	Harper et al. (2000)
Micromet.	Spring	278-298	11-34	Harper et al. (2000)
Micromet.	Summer	574	42-59	Harper and Sharpe (1998)
Micromet.	Winter	538	14-33	Harper and Sharpe (1998)
Micromet.	Spring	741	14-42	Harper and Sharpe (1998)
Micromet.	Summer	193	7.0-20	Harper and Sharpe (1998)
Micromet.	Winter	183	14-22	Harper and Sharpe (1998)
Micromet.	Spring	227	7.2-16	Harper and Sharpe (1998)
Chamber	Summer	587-695	145	Aneja et al. (2000)
Chamber	Fall	599-715	30	Aneja et al. (2000)
Chamber	Winter	580-727	11	Aneja et al. (2000)
Chamber	Spring	540-720	63	Aneja et al. (2000)
TG OP-FTIR	May	-	93-305	Todd et al. (2001)
TG OP-FTIR	November	-	20-169	Todd et al. (2001)
Chamber	September	101-110	0.44-2.7	Aneja et al. (2001)
Chamber	November	288-311	0.04-0.14	Aneja et al. (2001)
Chamber	November	350	0.17-0.62	Aneja et al. (2001)
Chamber	Feb/March	543-560	0.35-2.6	Aneja et al. (2001)
Chamber	March	709-909	0.32-1.2	Aneja et al. (2001)
Chamber	April-July	978-1143	319	Heber et al. (2001)
Chamber	May-July	326-387	48	Heber et al. (2001)

[a] Micromet. = micrometeorological; TG OP-FTIR = tracer gas open path fourier transform infrared spectroscopy; Chamber = dynamic flow through chamber.
[b] TAN = total ammoniacal nitrogen.
SOURCE: Data derived from Tables 9 and 10, Arogo et al., 2001

regarding the accuracy of emissions calculated for individual AFOs. In addition, large uncertainties associated with emission rates for the principal components of a manure management system reduce the probability of documenting success in the application of emission reduction technologies.

As a first approximation, estimates of the variance associated with emission rates, such as those in Table 2-2 can be obtained using normal statistical procedures. If estimates of the variance are included in published reports, then the variance associated with the derived emission factor can be calculated by using well-known formulas for the propagation of error (Beers, 1957), and assuming no significant autocorrelation between sequential observations. As noted earlier, emissions from AFOs are most likely parts of time series with autocorrelation between observations, especially those taken over relatively short periods of time (hours or days). The presence of autocorrelation within a data set means that calculated values for the variance of the sample mean using standard statistical procedures will be biased low, and that the overall uncertainty for a derived emission factor will be underestimated.

When values of the variance associated with emission rates are not included in the published literature, very rough approximations of the population variance can be obtained from the range of reported values (Natrella, 1963; Deming, 1966). For example, if it is assumed that the data follow a normal distribution and the reported range in emission rates encompasses 95 percent of the sample population, the estimate of the population standard deviation (σ) is

$$\sigma = \frac{(\text{maximum} - \text{minimum})}{4} \qquad \text{(Eq. 2-3)}$$

Values for the denominator in Equation 2-3 range from 3.5 (random) to 4.9 (triangular) for other assumed shapes of the data distribution (Natrella, 1963). For the purposes of this report, the data are assumed to follow a normal distribution.

Applying Equation 2-3 to the data in Table 2-2, and assuming that each population mean is equal to the average of the minimum and maximum values, we find percent CV values ranging from 8.4 to 42.6 for the individual combinations of measurement method and measurement period, with a mean (for 17 entries) of about 25 percent. This is similar to values noted earlier for field measurements (Groot Koerkamp et al., 1998; Doorn et al., 2002), and reinforces the argument that the uncertainty associated with published values of emission rates (or flux) cannot be ignored when deriving emission factors. These calculations illustrate that at a minimum, a derived emission factor for NH_3 emissions for a single AFO based on the data in Table 2-2 will probably have an associated CV of at least 25 percent. This is a minimum estimate because our calculations using the data in Table 2-2 are based only on the within-study variance.

The approach for estimating uncertainty represented by Equation 2-3 can provide only a rough estimate of the standard deviation of the sample population. If the reported range in emission rates represents a limited number of observations, then the assumption that the range encompasses 95 percent of the possible observable values is less likely to be true. Proper characterization of the uncertainty associated with emissions in the published literature, therefore, also requires knowledge of the number of observations. This is especially important when averaging values for derived emission factors as is done by EPA (2001a). Simple averaging implies equality in the uncertainties associated with the emission rates used to determine emission factors. In reality, the actual numbers of observations associated with reported values in the published literature vary substantially among investigators, requiring serious consideration of weighted averaging as a more valid means of calculating emission factors. Developing a weighting protocol will require examination of the experimental design employed for each set of emissions data considered, determining the most likely sources of variation in the reported values, and considering whether the experimental design gathered sufficient data to obtain realistic estimates of this variation. Weighted averaging is not considered by EPA (2001a).

The model farm construct proposed by EPA (2001a) attempts to reduce the uncertainty in deriving emission factors for individual AFOs by subdividing the overall AFO population according to the manure management systems used. Subdivision of large sample populations into smaller subsets is an acceptable procedure to reduce uncertainty (i.e., improve sample quality). The measurement of emissions from an individual AFO (or component of an individual AFO) will necessarily be interpreted as being representative of all AFOs in a defined subset of the larger sample population. However, further subdivision of the sample population also increases the need for data in terms of emission rates and emission factors. This approach must necessarily reach a point of diminishing returns.

Emission rate measurements obtained on two AFOs using the same management schemes for animal housing and manure handling will likely not be the same. To include both operations in the same sample AFO population will therefore require the overall uncertainty in the emission factor to be increased to allow both to be part of the same statistical population. Attempting to use only a mean value for a sample population to characterize an individual member of that population must necessarily have a large degree of uncertainty associated with it. To decrease this uncertainty, specific information concerning the individual member of the sample population to be characterized must be included in deriving the estimated value. This necessarily will increase the complexity of the model used to describe individual members of the population and therefore the size of the database required to accomplish the desired goal.

In summary, an example has been given of how the statistical uncertainty in emissions measurements and emissions factors can be characterized in the scientific literature, provided sufficient information is available in published reports. The example speaks solely to the issue of precision and cannot address the question of accuracy (systematic bias) of the reported values. However, issues concerning systematic bias have been addressed elsewhere in this chapter. Failure of investigators to note the degree

DETERMINING EMISSION FACTORS

of uncertainty associated with their reported values for emission rates may be a reflection of the limited number of observations upon which their reported values are based. Equal weighting should not be given to reported emission rates and derived emission factors when the actual number of observations on which these reported values are based differs significantly among investigators. All other things being equal, reported values for emissions based on a relatively large number of observations should be given greater weight than those derived from relatively few observations.

As presented in this chapter, a wide range of factors can influence air emissions of gases, PM, and other substances from AFOs. Combinations of these factors that will be most useful in pursuing regulatory goals will depend on research-based information about the strength of the relationship between each combination of factors and the rate of emission of a particular pollutant.

> **Finding 5: Reasonably accurate estimates of air emissions from AFOs at the individual farm level require defined relationships between air emissions and various factors. Depending on the character of the AFOs in question, these factors may include animal types, nutrient inputs, manure handling practices, output of animal products, management of feeding operations, confinement conditions, physical characteristics of the site, and climate and weather conditions.**

3

Models For Estimating Emissions

This chapter examines the approach for estimating air emissions used in the draft report to the EPA, *Air Emissions from Animal Feeding Operations* (EPA, 2001a), problems with the approach, and issues that must be addressed in getting supportable estimates. Model farms are used to define hypothesized relationships between air emissions and selected characteristics of various kinds of large operations that produce a large proportion of the livestock animals marketed in the United States.

Some variation of the model or average feeding operation appears to the committee as necessary as a basis for estimating air emissions from individual farms. The issue to be faced is finding the combination of characteristics of feeding operations that can be used to estimate air emissions with desired levels of accuracy and at reasonable costs. In the sections that follow, the committee assesses (1) the viability of the particular model farm approach used in the EPA draft report; (2) whether it can be improved using available data; (3) alternative approaches based on model farms constructs; (4) ways to characterize the substances emitted and the components of manure to be estimated; and (5) mitigation technologies and management practices in addition to those identified in the EPA draft report.

EPA MODEL FARM CONSTRUCT

Are the emission estimation approaches described in the EPA/OAR summary document, *Air Emissions from Animal Feeding Operations,* **appropriate?** The goal of EPA (2001a) was "to develop a method for estimating [air] emissions at the individual farm level that reflects the different animal production methods that are commonly used at commercial scale operations." The approach is intended to provide estimates of total annual air emissions from animal feeding operations (AFOs) for defined geographic areas by kind of animal and manure handling practices for each of eight kinds of emissions. It does this with a model farm construct that provides estimates of average annual emissions per animal unit (AU) for twenty-three model farms (two for beef, eight for dairy, two for poultry-broilers, two for poultry-layers, two for poultry-turkeys, five for swine, and two for veal; Appendix D). Each model is defined by three variable elements that describe manure management practices for typical large AFOs: (1) confinement and manure collection system, (2) manure management system, and (3) land application. The manure management system was further subdivided into solids separation and manure storage activities. Insofar as combinations of these elements are regionally distinctive, the model farms also reflect regional variations in air emissions.

Model farms, as used by EPA (2001a), are a useful device for aggregating emission rates across diverse sets of AFOs. A model farm can be used to represent the *average* emissions across some geographic area over some period of time per unit capacity of a class of farms (e.g., all pig farms in the United States that use an enclosed house with pit recharge and irrigation of supernatant onto forage land; model farm S2 in Appendix D).

The applicability and use of a model farm construct of the kind used by EPA (2001a) depends on:

- defining models in which the dependent variable, the amount of an air emission per unit of time, is closely related to independent variables that accurately depict real feeding operations, and that can explain a substantial share of the variation in the dependent variable;
- providing accurate estimates of the relationship between the dependent and independent variables in the model farm construct; and
- having estimates of the relationship between dependent and independent variables that clearly distinguish among the kinds of AFOs being modeled.

A critical data requirement for estimating the appropriate emission factors is a statistically representative survey of emissions from the class of AFOs over several iterations of the time period to be represented. The size of the sample required to estimate the mean emission rate with a given degree of statistical significance increases with the variability of the factor to be measured (dependent variable) across the set of variables (independent variables) that affect it. Independent variables that have been discussed include animal type and age, diet, local climate, building type, land application method, and management. To the extent that some of these variables change over time (e.g., trends in farm organization, location, practices, and technology), updating of estimates and estimates of trends may be required.

The model farm construct is represented by Equation 3-1:

$$E = \Sigma\,(w_i \times e_i) \qquad \text{(Eq. 3-1)}$$

in which the emission (E) of a particular pollutant from an AFO during a period of time is the product of the emission (e_i) from each unit on the model farm and the number of units (w_i) of that type, summed over the farm.

One use of model farms is to predict emissions and local effects for a single AFO or cluster of AFOs in a small area. This is a different use from that described by EPA (2001a) and requires a detailed model of the effects of selected variables on the rates of emissions and their downwind concentrations. An example of this type of model is an odor dispersion model that predicts odor intensity and frequency at various locations, given information on odor sources and local meteorological conditions. More data (perhaps hourly) and statistical analyses of the relationships between various explanatory variables and pollutant concentrations or impacts are required. A starting point for classifying types of data needed by emission type and intended use of its emission factor is shown in Table 3-1.

The committee believes that EPA (2001a) fails to meet these standards. It does not provide a methodology to adequately determine air emissions from AFOs because both the model farm construct and the data are inadequate. Concerning the former, the model farm construct used by EPA (2001a) cannot be supported for estimating air emissions from an individual AFO. There is a great deal of variability among AFOs that cannot be accounted for using this approach. (See Finding 7.) In particular, additional factors not included in the EPA model that affect emissions include animal feeding and management; animal productivity; housing, including ventilation rate and confinement area; use of abatement strategies such as sprinklers to decrease dust; and physical characteristics of the site such as soil type and whether the facility is roofed. In addition, emissions are likely to differ for different climatic (long-term) and weather (short-term) conditions including temperature, wind, and humidity. Thus, accurately predicting emissions on individual AFOs would require determination of emission factors that reflect these characteristics. Accurate estimates of these emission factors would require sampling hundreds of AFOs representing different management and meteorological conditions. The cost of accurately measuring emissions on the number of AFOs (i.e., thousands) that would be needed to replicate all common situations would be very high.

TABLE 3-1. Classification of Emissions by Likely Intended Use of Emission Factors

Emission Type	Intended Use of Emission Factors		
	Regional Annual Inventory	Local Seasonal Ambient Effects	Local Transient (Hourly) Effects
NH_3	X	X	X
CH_4	X		
VOC	X	X	X
PM		X	X
H_2S		X	X
N_2O	X		
NO	X		
Odor		X	X

NOTE: CH_4 = methane; H_2S = hydrogen sulfide; NH_3 = ammonia; NO = nitric oxide; N_2O = nitrous oxide; PM = particulate matter; VOC = volatile organic compounds

More specifically, improvements in the model farm construct are needed for both discrete variables (e.g., management, confinement conditions, location) and continuous variables (e.g., nutrient input, productivity, meteorology). Concerns about quality of data (use of non-peer reviewed data), lack of data, inappropriate use of data, and representativeness of the data were discussed in Chapter 2.

Finding 6: The model farm construct as described in EPA (2001a) cannot be supported because of weaknesses in the data needed to implement it.

Finding 7: The model farm construct used by EPA (2001a) cannot be supported for estimating either the annual amounts or the temporal distributions of air emissions on an individual farm, subregional, or regional basis because the way in which it characterizes feeding operations is inadequate.

INDUSTRY CHARACTERIZATION

How should industry characteristics and emission mitigation techniques be characterized? This question asks for suggestions to improve the approach described in the EPA draft report. The committee has discussed several inadequacies in the EPA approach. In the next section, an alternative approach suggested by the committee is discussed in some detail. Rather than discuss possible improvements in estimating air emissions using the EPA approach and the use of possible emissions mitigation techniques based on the EPA estimates at this time, these issues are being left to the final report.

Mitigation of air emissions based on best management practices, including those under comprehensive nutrient management plans (CNMPs) is an option already being used in various places. Although the effectiveness of the best management practices approaches is not wholly clear to the committee at this time, especially in the absence of research-based data on mass balance approaches, those practices that are already being used provide a basis for action until better information is available.

PROCESS-BASED MODEL FARM APPROACH

Should model farms be used to represent the industry? If so, how? What substances should be characterized and how can inherent fluctuations be accounted for? What components of manure should be included in the estimation approaches (e.g., nitrogen, sulfur, and volatile solids)? The committee has discussed using a process-based model farm approach to predict emissions on individual AFOs. A process-based approach would use mathematical modeling and experimental data to simulate conversion and transfer of reactants and products through the farm enterprise (Denmead, 1997; Jarvis, 1997). This alternative to EPA's model farm approach (EPA, 2001a) would involve analysis of the farm system through study of its component parts. Rather than simply add the emissions observed from each farm element, a mathematical model would be used to represent the interactions between the system components (see Figure 3-1 for a representation of an animal production enterprise). Development of a process-based model does not obviate the need for data collection, but it enables the use of data representing only part of the farm system and will help identify gaps in the existing literature.

For many pollutants (e.g., NH_3 [ammonia], H_2S [hydrogen sulfide], and CH_4 [methane]), the quantity of emissions is likely to be proportional to the amount of material (substrate) from which the pollutant is derived. For example, the amount of NH_3 emitted from a manure slurry is expected to be proportional to the amount of nitrogen in the manure (Muck and Steenhuis, 1982). With a compartmental modeling approach (Jarvis, 1993; Dou et al., 1996) and an assumed steady state, nitrogen in manure can be determined as intake nitrogen minus animal product nitrogen. Further, NH_3 volatilization from manure during collection can be estimated as a fraction of manure nitrogen produced. The NH_3 volatilized from storage can be represented as a fraction of nitrogen remaining after collection, and NH_3 volatilization during field

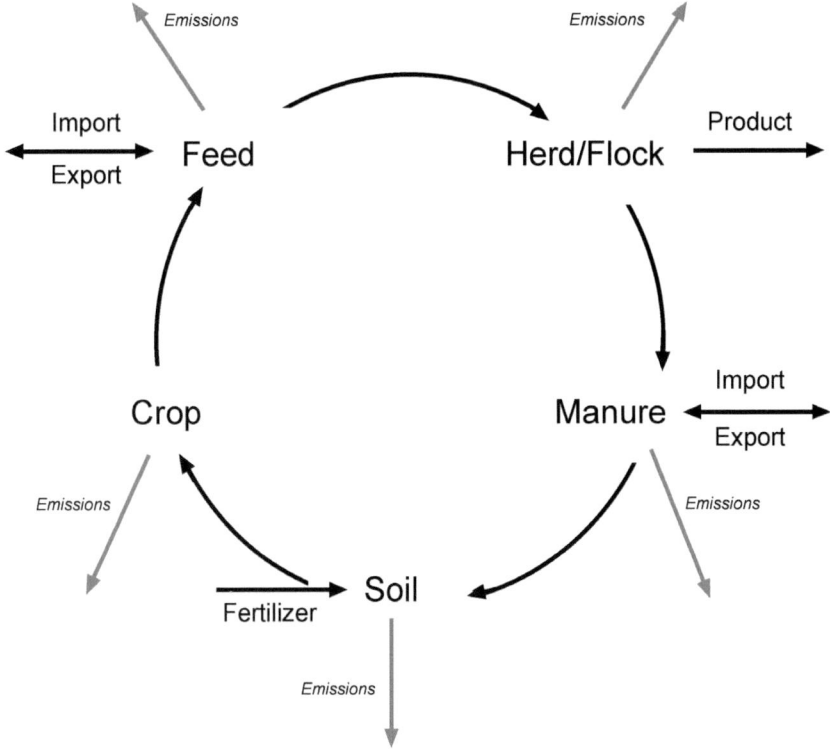

FIGURE 3-1. A process-based model of emissions from an animal feeding operation.

application can be represented as a fraction of nitrogen applied (Denmead, 1997).

There appears to be a disappearance of nitrogen from manure storage or from soil in the form of harmless nitrogen gas (N_2) (Thompson et al., 1987). Thus, the ratio of NH_3 to N_2 emissions would have to be determined under different animal management and meteorological conditions. There is little research, and even less agreement, as to what proportion of nitrogen is lost from various types of manure storage as NH_3 or N_2 (Harper et al., 2000). Nonetheless, much of the variation in emissions from AFOs, such as that from feeding and animal management, can be accounted for by predicting the effect on manure nitrogen production (Kohn et al., 1997). Other factors, such as climate and management conditions that affect partitioning of nitrogen from storage and land application, would be accounted for in the models as knowledge of how they influence processes becomes available.

Development of a process-based model of emissions will require a large amount of data, but the number of farms that would have to be represented would be reduced. Using a strictly empirical approach to estimate emissions would require measurements on farms representing the full diversity of agriculture in the United States. For example, emissions would be determined on farms using different combinations of animal, feed, and manure and crop management. With the process-based approach, emissions would be determined from different farm components and mathematical calculations used to determine emissions for different combinations of components. Furthermore, more data may be available to develop estimates of emissions from farm components than are available for whole-farm emissions.

Different models would be needed to fit different objectives for the prediction estimates. Prediction of annual rates of emissions would require understanding relationships in a more aggregated way than prediction of potential short-term effects. When considering the acute health effects of emissions for nearby residents, short-term potential emissions would be needed, and a dynamic process-based model to predict emissions on a daily or more frequent basis may be recommended. When considering long-term atmospheric emissions, an aggregated model on an annual time step may be adequate. If emission rates are needed to categorize farms that may potentially emit enough pollutants to warrant extra regulation, tabular values representing typical animal, crop, feed, and manure management might be adequate, and predictions for different situations could be calculated and reported in tables for rapid referral.

As with NH_3, emissions of other nitrogen-containing compounds can best be estimated as a fraction of excreted nitrogen to emittant (Müller et al., 1997). However, other factors such as soil compaction and oxygen and moisture content also contribute substantially to variations in NO (nitric oxide) and N_2O (nitrous oxide) emission processes (Li et al., 1992; Dendooven et al., 1996). The approach may involve modeling the ratio of $N_2O:N_2$ and factors that affect it, because generation rates of both gases are linked to rates of nitrification and denitrification (Abbasi et al., 1997), and management may be able to shift reactions to favor the more benign product, N_2 (Dendooven et al., 1996).

The emissions of H_2S are likely to be a function of the amount of sulfur delivered to anaerobic manure storage; manure sulfur will be equivalent to the sulfur in feed and water minus the sulfur in animal products (including growth). Whereas most sulfur will be converted to H_2S by microorganisms under anaerobic conditions, the rate of H_2S volatilization will depend on pH and other factors.

Methane emissions from a ruminant animal are proportional to the carbohydrate content of its diet, with additional effects caused by forage to concentrate ratio and the use of ionophores (Johnson and Johnson, 1995). CH_4 emissions from manure storage could also be expressed as a fraction of the carbon delivered to storage, including undigested feed and bedding material.

Particulate matter (PM) emissions occur primarily from feeding and housing. Quantifying the total feed used may explain some variation in PM emissions from feeding, but a different means is necessary to estimate emissions from housing.

Once a farm has been identified as a potentially high risk, actual farm-specific data such as feed amounts and manure analysis could be used in the models to more accurately predict emissions. This approach would reward producers who can document reducing inputs of substrates for emissions (and presumably outputs of emissions) and provide more of a performance standard rather than a prescriptive regulation, but without the cost and uncertainty involved in measuring actual emissions.

Finally, certain sectors of the animal enterprise are likely to be more important for some emissions than for others. Development of a process-based model would enable system analysis and simulation for determining critical control points for emissions (Kohn et al., 1997). It would also highlight fruitful research areas, and identify knowledge gaps that need to be filled in order to improve understanding of farm processes.

Finding 8: A process-based model farm approach that incorporates "mass balance" constraints for some of the emitted substances of concern, in conjunction with estimated emission factors for other substances, may be a useful alternative to the model farm construct defined by EPA (2001a). The committee plans to explore issues associated with these two approaches more fully in its final report.

MITIGATION TECHNOLOGIES AND BEST MANAGEMENT PRACTICES

What additional emission mitigation technologies and management practices should be considered? Previously, research on emission mitigation technologies and management practices for AFOs has been limited. However, more research in these areas is anticipated over the next several years. An exhaustive list of potential technologies would be difficult to produce, so the committee has highlighted several ongoing research efforts around the country to introduce some of the technologies and management practices that may prove useful in decreasing air emissions from AFOs. Undoubtedly there are technologies not discussed here that may prove to be as good as those listed. Lack of inclusion should not be construed as dismissing their potential. The committee will explore mitigation technologies and best management practices more thoroughly in its final report.

Animal Feeding Strategies

Feeding livestock closer to their nutrient requirements may result in decreased nutrient content of manure and subsequent decreases in emissions of certain pollutants (e.g., NH_3, H_2S). Nutrient requirements of livestock species have been determined and well documented (National Research Council, 1994, 1998, 2000, 2001).

Several approaches for decreasing manure nitrogen production are available. Increasing production of salable food products (meat, milk, and eggs) per animal decreases the number of animals required to fill the market demand for those products. The animal's requirements can be divided into needs for maintenance (maintaining basal metabolism) and production (National Research Council, 1994, 1998, 2000, 2001). By meeting maintenance requirements while increasing production, nitrogen emissions from manure are decreased. Dunlap et al. (2000) showed that increasing milk production of dairy cows by administering growth hormone, increasing photoperiod using artificial lighting, and milking three times daily instead of two can decrease manure nitrogen by 16 percent for a given amount of milk produced.

In addition to increasing production per animal, nitrogen excretion to manure can be decreased by feeding at a level closer to the animal's requirements. Grouping animals with similar requirements makes it possible to feed them more precisely to meet their requirements with the same diet. For example, broilers are already separated by age, but greater homogeneity may be obtained by separating by sex as well (Fritz et al., 1969). Also, feeding broilers four different diets over the course of their lifespan, rather than the standard three, results in decreased nutrient inputs by 10 percent (Dhandu, 2001). Grouping dairy cows into separate production groups on a farm decreases nitrogen excretion by 6 percent compared to feeding all lactating cows the same ration (St-Pierre and Thraen, 1999).

Of all current practices, feeding amino acid supplements has had the greatest impact on decreasing nitrogen excretion in manure. Animals require a specific profile of amino acids for optimal production, but most feeds do not provide that profile. When balancing the diets of animals, corn and legumes are typically mixed to provide a complementary set of amino acids. Corn is high in methionine but low in lysine, while legumes are high in lysine and low in methionine. Synthetic amino acid supplements can be used to further decrease protein feeding without sacrificing production or health. Sutton et al. (1996) showed that for growing pigs, corn and soybean meal diets supplemented with lysine, tryptophan, threonine, and methionine decreased NH_3 and total nitrogen in freshly excreted manure by 28 percent. Using amino acids that are protected from degradation in the rumen of cattle have been shown to decrease nitrogen excretion from dairy cattle by as much as 26 percent (Dinn et al., 1998).

Manure Handling and Treatment

Once excreted from animals, manure naturally undergoes microbial decomposition, usually anaerobic. A number of inorganic gases and organic compounds are produced during the decomposition process. Manure handling and treatment can have a great influence on the physical, chemical, and biological properties of manure and consequently on the emissions of air quality concern. Solid and liquid manure are handled differently on AFOs. There are many treatment technologies available that could play important roles in emission mitigation. However, the effectiveness of most of them is not well quantified. Standard protocols for evaluating the air quality impact of different manure handling and treatment technologies must be developed. Some technologies may reduce emissions of certain gases or compounds but increase emissions of others. Treatment technologies have to be analyzed with clear objectives as to what emissions are to be mitigated. Two recent literature summaries (Lorimor et al., 2001; Sweeten et al., 2001) reviewed various animal manure handling and treatment technologies that have been used on AFOs or extensively researched. A whole-farm approach needs to be taken when evaluating emission mitigation technologies. Knowledge of animal manure distribution on AFOs and emission source characterization from individual sources (such as animal houses, feedlots, manure storage, and land application) is important for quantifying potential emission mitigation effects of new technologies. Recently, a project was initiated by United States Department of Agriculture Natural Resources Conservation Service to identify and evaluate the emerging animal manure treatment technologies that are most likely to be used by animal producers in the next five to ten years. The project was led by Iowa State University and supported by a four-member advisory board. A preliminary list of manure handling and treatment technologies that have been identified and have relevance to air emissions includes: storage covers, anaerobic digestion, aeration, solid-liquid separation, composting, and chemical treatment for pH control (Melvin, personal communication, 2002).

The potential air quality impacts of these manure treatment technologies will be analyzed in the committee's final report based on the published information, with recommendations for further research and development.

North Carolina

On July 25, 2000, Smithfield Foods, Inc., entered into a voluntary agreement with the Attorney General of North Carolina to provide resources for an effort to develop innovative technologies that are determined to be technically, operationally, and economically feasible for the treatment and management of swine wastes (Williams, 2001). Performance standards, along with comprehensive analyses of odor, NH_3, and pathogen emissions, as well as economic analyses, are required for each technology. Currently, 16 systems are being studied:

- psychrophilic (unheated and unmixed) ambient temperature anaerobic digester, energy recovery, greenhouse vegetable production;
- thermophilic (high-temperature) anaerobic digester energy recovery;
- solids separation-constructed wetlands;
- sequencing batch reactor;
- upflow biofiltration;
- solids separation, nitrification-denitrification, soluble phosphorus removal, solids processing;
- belt manure removal and gasification to thermally convert dry manure to a combustible gas stream for liquid fuel recovery;
- ultrasonic plasma resonator;
- manure solids conversion to insect biomass (black soldier fly larvae) for value-added processing into animal feed protein meal and oil;
- solids separation-reciprocating water technology;
- microturbine cogeneration for energy recovery;
- belt system for manure removal;
- high-rate second-generation totally enclosed Bion system for manure slurry treatment and biosolids recovery;
- combined in-ground ambient digester with permeable cover or aerobic blanket, BioKinetic aeration process for nitrification-denitrification, in-ground mesophilic anaerobic digester;
- dewatering, drying, desalinization; and
- solids separation-gasification for energy and ash recovery centralized system.

California

The State of California recently awarded a $5 million grant (matched by $4.8 million in federal funds) to develop a centralized waste processing facility in Chino, California. Effects of this centralized treatment have not yet been evaluated. The State also provided $10 million as cost sharing for dairy farmers to build anaerobic digesters. So far there are more than 30 applications from dairy farmers interested in participating in the cost-sharing program. Research is being carried out at the University of California, Davis on alternative manure treatment technologies such as solid-liquid separation, aeration and anaerobic digestion.

USDA Agricultural Research Service Air Quality National Program

In January 2000, the Agricultural Research Service (ARS) met with stakeholders in Sacramento, California, to explore air quality problems associated with agriculture (USDA, 2002). This meeting was the first step in developing a list of high-priority research needs and a research program to address those needs. The Agricultural Air Quality Task Force had previously provided the Secretary of Agriculture with a list of research needs. EPA has been actively seeking ARS research in several agricultural air quality topic areas. ARS research in agricultural air quality is organized into five categories:

1. particulate emissions,
2. ammonia and ammonium emissions,
3. malodorous compounds,
4. ozone impacts, and
5. pesticides and other synthetic organic chemicals.

4

Assessing the Effectiveness of Emission Mitigation Techniques and Best Management Practices

What criteria, including capital costs, operating costs, and technical feasibility, are needed to develop and assess the effectiveness of emission mitigation techniques and best management practices? The criteria for evaluating emission mitigation techniques should provide sufficient information to analyze probable societal effects of proposed changes in policy and regulations. The relevant effects are the direct biological and health effects of the emissions themselves and their related economic impacts. It is imperative that a comprehensive holistic approach be adopted. Farms are composed of several interrelated components spread over a significant geographic area. The approach to evaluating mitigation techniques must clearly identify the portion of the system being evaluated and measure changes in all of the material flows (rates and compositions of inputs, air emissions, and liquid and solid effluents) and economic inputs and outputs. The changes in inputs and outputs brought about by adoption of the technique must then be used to model effects elsewhere on the farm and beyond it. Failure to adopt a comprehensive approach risks ignoring increased air emissions elsewhere and having increased adverse environmental effects on land or water resources. A comprehensive evaluation should allow policy analysis that includes quantification and valuation of all the predictable effects on social welfare, including public health, the environment, and the economy. Extensive literature exists on analysis of costs and benefits of policy. Arrow et al. (1996) make the case for benefit-cost analysis. Examples of textbooks on the topic include Layard and Glaister (1994) and Boardman et al. (2001).

CRITERIA FOR EVALUATING EMISSIONS EFFECTS OF MITIGATION TECHNIQUES

Criteria for evaluating mitigation techniques emphasize information needs for policy analysis. These include both "on-farm," or primary, effects of changes in policy, incentives, and regulations, and "off-farm," or secondary, effects. The primary effects include changes in the composition and rates of emissions from farms subject to changes in policy. These farms may adopt mitigation techniques, decrease or cease production, begin or expand production, or otherwise modify production practices and management, all of which are likely to affect air emissions. Information needs for policy analysis also include those related to secondary effects, such as increased air emissions from trucks hauling manure greater distances as a result of changes in regulations.

Analysis of policy changes should, at a minimum, capture the following factors:

- effects of changes in land application of manure on groundwater and surface water quality;
- effects of the risk of occasional events, such as storms, and policy-related changes in emissions due to those events;
- changes in material flow and composition that can be used to analyze secondary effects.

For example, if a proposed change in policy requires impermeable covers on anaerobic treatment containments, then changes in the flow and composition of the supernatant and sludge leaving them must be measured, as well as changes in the rate and composition of direct air emissions from them. Changes in the flow and composition of effluents from the containment can then be used to analyze changes in air emissions and other effects occurring beyond the containment. In this example, such effects might include increased undesirable air emissions from open secondary storage containment, livestock buildings that use recycled containment supernatant for flushing, land on which supernatant and sludge are applied, and increased energy generation required to distribute the supernatant and sludge over a greater area.

Estimated changes in the composition and rate of air emissions resulting from a policy change can be evaluated using fate and transport models and their predicted changes in impacts on public health and the environment. This interim report does not address the accuracy or statistical validity of models that transform emissions estimates into predicted impacts on public health and environmental quality.

CRITERIA FOR EVALUATING ECONOMIC EFFECTS OF MITIGATION TECHNIQUES

The last two sections of this report are presented as an overview of the information needed for relevant economic analyses of the effects of changing policies, including regulations and incentives, on mitigating air emissions. Economics is the study of the optimal use of resources to maximize human welfare. A thorough economic analysis of a proposed change in policy requires quantification and valuation of all the public health and environmental effects of the change, as well as its immediate and long-term effects on wealth, income, and employment. This section examines criteria to evaluate immediate and long-term effects on wealth, income, and employment. Emphasis is on identifying financial and economic information to be collected to evaluate mitigation techniques. The environmental and public health effects, including both costs (negative effects) and benefits (positive effects), will be examined in more detail in the final report.

Wealth and income as measures of economic welfare (well-being) are usually described in terms of values determined by market transactions. Buyers' willingness to pay is matched with sellers' willingness to accept payment; this works well in setting market prices for many commonly traded market goods and services. It frequently does not work well in setting comparable values for goods and services that are not traded in ordinary markets, as is often the case for human and environmental health. It is commonly accepted that improvements in both are beneficial and are valued, but estimating this value in terms that can be compared with market-determined values is difficult.

Economists call these kinds of benefits and costs "externalities." They are recognized as being real, but their values are determined outside ordinary markets. Various ways of framing these values have been devised so that they (or proxies for them) can be weighed in decisions that also involve market values (National Research Council, 1999). For example, protocols for cost-benefit analyses for evaluating federal projects typically include guidance on handling externalities. This issue will be treated in the committee's final report, in which effects of air emissions on health and ecosystems will be discussed in more detail. Techniques for estimating benefits in the absence of direct market data include hedonic analysis (using changes in values of associated goods to estimate changes in the value of the good in question) and contingent valuation (using controlled consumer surveys to estimate values attributable to actions such as mitigating air emissions). The remainder of this chapter focuses on evaluations based on market-determined prices.

Changes in policies and regulations related to air emissions from livestock operations are likely to lead to changes in economic performance of affected farms. They may also result in changes in local, regional, and national economies. A thorough evaluation of the economic effects of these changes requires a detailed analysis. Criteria to evaluate air emission mitigation techniques must capture the material flow and economic effects on farms that adopt mitigation techniques. Predicted economic effects on farms adopting mitigation techniques must be sufficient for use in modeling off-farm economic effects as required for policy analysis.

Estimates of farm-level economic effects must be fully consistent with estimates of emissions and material flow effects to allow correct analysis of impacts of policy change. Accurate estimates of material flow and economic effects at the farm level can then serve as the basis for modeling the local, regional, and national economic effects of adoption of mitigation techniques. Attention in modeling local and regional economic effects should be given to the following:

- changes in the demand by livestock farms for goods and services;
- changes in the supply of manure or manure by-products for farmland application and other uses;
- changes in demand or supply of goods and services affected by air emissions from livestock farms; and
- associated direct and indirect effects on income, employment, investment, and tax base throughout the local and regional economies.

National economic effects that merit attention include:

- changes in prices and quantities of livestock produced;
- changes in imports and exports of livestock products;
- changes in the national supply and demand for goods and services related to livestock production and air emissions mitigation;
- the aggregate effect of changes in the regional income, employment, investment, and tax base; and
- the resulting changes in producer and consumer welfare.

The emphasis on measuring "changes" in the above protocol is important. The effects to be measured are those that derive from changes in farm practices, especially in response to changes in policies and regulations affecting air emissions. These changes are the well-known "marginal," or incremental, changes that are the basis for most economic analyses.

Although the emphasis for economic analysis is on marginal changes, effective analysis requires a clear understanding of the basic operations and economics of the farm enterprises being addressed. Thus, criteria for evaluating farm-level economic effects should capture the main economic factors that affect farm operations: costs, revenues, financial status, limited resource feasibility, and exposure to risk of substantial financial loss (liability). Limited resource feasibility refers to the ability of the farm to implement new techniques given limited quantities of available labor, land, and management. Each of these factors requires some discussion. (The investment and cost analysis methods described below are consistent with those described in the Environmental Protection Agency [EPA] *Air Pollution Control Cost Manual* [EPA, 2001b]). Textbooks on farm management provide farm specific methods for calculating investment, cost, revenue and profit, as well as farm enterprise feasibility. Examples include James and Eberle (2000), Kay and Edwards (1999), and Boehlje and Eidman (1984).

Farm-level costs include capital costs (those associated with initial investment), operating costs (those that recur annually), and occasional costs (those that occur occasionally in the life of the project, such as sludge removal from a containment every 5 or 10 years).

Capital costs include the costs of:

- purchasing and installing equipment;
- designing and constructing structures and land modifications;
- establishing pastures or groundcover that will last more than one year;
- installing new utility connections;
- obtaining permits, leases and rentals used in construction;
- interest accrued on capital committed to construction; and
- the value of unpaid inputs, such as the owner's labor, management, equipment, and capital.

Initial investment may be reduced (or increased) by the net salvage value (net closure or removal cost) of the facility at the end of its useful life. Initial investment may also be reduced by the amount of cost share or other subsidy received. Initial investment is converted to *annualized capital costs* by amortizing it over the expected or typical useful life of the facility, using an appropriate interest rate. The interest rate should reflect the owner's cost of borrowing money over the amortization period. Criteria for mitigation technique evaluation should ensure that component description, type and capacity, expected life, price, and installation cost are reported.

Operating costs include labor and management (hours, wages, and benefits), fuel, electricity, supplies (additives, lubricants, filters, etc.), repairs and maintenance, rentals and leases, royalties, permit fees, fines, custom and professional service costs, insurance and taxes, interest on operating capital, reduction in the value of assets or inventory, the value of unpaid goods and services contributed by the owner or others, and any other expenses incurred in owning and operating the facility. Criteria for mitigation technique evaluation should capture the quantity, quality or type, and price of each input consumed. Operating costs may be reduced by any cost sharing or other subsidy received to offset such costs.

Occasional costs may include significant equipment overhaul, reseeding of groundcover, sludge removal from containment, and other costs that occur less frequently than annually. Evaluation criteria should capture the expected or typical timing of such costs (e.g. every fifth year), the cost per occurrence (including quantities and prices where appropriate), and any other relevant factors. An important consideration for occasional costs is whether the cost estimates are in current dollars or have been adjusted to allow for inflation. If all costs are in current dollars, then a 'real' discount rate (typically 3 or 4 percent) can be used to deflate a series of occasional costs to their net present value at the time of the initial investment. If occasional costs have been adjusted to include inflation, then a 'nominal' interest rate (e.g. 7 to 9 percent) can be used to deflate the cost series. That net present value can then be amortized over the life of the facility, similarly to initial investment, to produce an *annualized occasional cost* estimate.

Revenues include cash received from the sale of goods or services, an increase in the value of assets and inventory, savings (e.g., reduced costs of fertilizer or electricity) realized elsewhere in the operation, and any other effects that represent an addition to the wealth of the operator. Evaluation criteria should capture the annual value of revenue, including the quantity, quality or type, and net price received from each source of revenue. Where revenue is occasional (e.g., the fertilizer value of land-applied sludge when sludge is removed from containment), the method described in the previous paragraph can be used to discount to net present value and annualize through amortization. Revenue may be increased by subsidies received that were not used to reduce initial investment or operating costs.

Evaluation criteria must capture or allow capture of effects on the *financial status* of the livestock enterprise and the farm. Financial status includes the value of debts compared to the value of assets, the ability to borrow money, cash flow (cash receipts versus cash outflow) and debt service capacity (ability to make scheduled debt payments), and profit (annual value of revenues versus costs). New investments in mitigation techniques can have undesirable effects on financial status because they may require new borrowing for a facility that has little or no resale value (no value as security for debt) and may introduce new costs with little or no revenue. Farmers may choose to close or sell their livestock operation if they are unable to borrow money to install required mitigation techniques, if the new costs would leave them unable to make scheduled debt payments, or if they are no longer able to generate a profit. The entire farm may be forced into bankruptcy if the change in financial status of the livestock operation reduces the financial status of the farm to an infeasible point. Financial status varies widely across farms, as does the relative financial importance of the livestock enterprise, so evaluation criteria for a new mitigation technology should capture its *marginal* impact on financial status (new capital required, new effects on debt versus assets, new effects on cash flow and debt service capacity, new effects on profit). This point underscores the importance of capturing both economic effects and effects on material flows and concentrations, since changes in material flows can also affect the economic viability of other enterprises on the farm.

Evaluation criteria should allow capture of effects on farm *limited resource feasibility*. Marginal changes in the required quantity and type of labor, management, and land should be estimated. Where these resources can be acquired easily in local markets, it may be sufficient to account for them as costs. However, where additional land or specialized labor or management is difficult to obtain, the farm may not be able to adopt the mitigation technique. This is important where livestock operations exist in clusters. The aggregate effect of a new regulation in causing many farms to seek to acquire a scarce resource may be quite different from its effect on a single isolated farm.

Evaluation criteria should allow capture of *exposure to risk of substantial financial losses (liability)*. Potential sources of new exposure to risk include those inherent in the mitigation techniques and the policy, such as major fines for occasional failure of the technique. Potential sources of new risk may also include increased threat of livestock or worker illness due to altered material flows on the farm. Criteria to evaluate risk may determine the effects of severe weather (wind, precipitation, floods, temperature), power outages, absence of workers, equipment failure, upsets of biological systems, and any other occasional event that could adversely affect the technique or the operation of the farm.

PARTIAL BUDGETING OR SELECTED COST AND RETURNS ESTIMATION

The primary method for evaluating the farm-level economic effects of adoption of an emission mitigation technique is *selected investment, costs and returns estimation*. This involves establishing a description of the mitigation technique and its component parts and activities, and a list of its direct effects on the livestock enterprise. A schematic showing the material flows and concentrations affected, as well as the goods and services required, is useful. A survey of farms using the technique is necessary to statistically determine their material flows, investment, costs, and revenues directly attributable to the mitigation technique.

Where the marginal impacts of the technique are difficult to determine, a survey of similar farms not using the technique may be needed to establish a basis for comparison. The term "selected cost and returns estimation" is used to emphasize that the analysis is focused on the mitigation technique rather than on the entire livestock operation or the entire farm. A limitation of this approach is that the researcher may omit items from the "selected" list and thereby underestimate or overestimate effects. A benefit of this approach is that it is less costly and less complicated than a whole-farm approach to cost estimation.

Where farms have yet to adopt a mitigation technique or where researchers seek to extrapolate from limited survey data, a *partial budgeting* approach can be used. Instead of relying on survey data for selected cost and returns estimates, the partial budgeting approach models initial investment and costs and returns using quantities and prices from secondary sources. The accuracy of predicted effects is dependent on the accuracy of prices and quantities used in the model, as well as its completeness. (For further exposition of partial budgeting methods, see Chapter 11 of Kay and Edwards, 1999.)

Problems in economic estimation based on partial budgeting are exacerbated when researchers must extrapolate. Examples of extrapolation (ranging from least calibrated to somewhat calibrated) include extrapolation from bench or pilot scale to full scale; from single full-scale prototype to multiple-farm implementation; and from farms in one region to multi-region implementation. Researchers and writers are obligated to caution readers about the degree of accuracy underlying extrapolated numbers. Regional differences can be partially accounted for in models by including critical design factors that are known to vary among regions. Similarly, differences in material flows, investments, costs, and revenues among farms of different types and sizes can be approximated by including known critical design parameters and equations.

OTHER CONSIDERATIONS IN EVALUATION OF MITIGATION TECHNIQUES

Surveys of selected investment, costs, and revenues of mitigation techniques can establish estimates of the range or variability of economic effects across farms. Effects are likely to vary because of farm-specific factors such as topography, soil type and crop production capacity, proximity to neighbors, proximity to environmentally sensitive sites, and so forth. Effects may also vary because of differences in the design or implementation of mitigation techniques. A selected sampling design may be used to establish a range of possible effects, while a larger randomized sampling design may provide estimates of variance in effects. Knowledge of the range or variance of effects for a farm of a given size, type, or region can substantially improve policy analysis.

Reporting the livestock capacity at each farm being treated by the mitigation technique surveyed is critical to extrapolating results. In partial budgeting applications, the type and number of livestock are critical inputs. Typical units or inputs for surveys and budgeting include the type of animal, number of head, stage of production, and steady-state live weight in inventory; other input may include the area of the feedlot or livestock building to be treated. Estimates of investment, cost, and revenue can then be reported with any of the physical input values as denominator.

Comprehensive analysis of prospective policy change requires a systems approach that captures direct and indirect effects. Criteria to evaluate air emission mitigation techniques should produce sufficient information to predict all relevant effects at the individual farm level, as well as at local, regional, and national levels.

Beyond the scope of the interim report but to be addressed in the final report is a broader discussion of the economics of policy change with respect to air emissions from livestock operations. Among the issues to be considered are the following:

- comparative response of farm managers to incentives versus regulations,
- the potential for value-added products from livestock manure and the associated potential to reduce waste and emissions,
- a consideration in the policy analysis of market structure including vertical integration,
- an expanded discussion of benefits estimation, and
- the analytical implications of global competitiveness.

REFERENCES

Abbasi, M.K., Z. Shah, and W.A. Adams. 1997. Concurrent nitrification and denitrification in compacted grassland soil. P. 47-54 in Gaseous Nitrogen Emissions from Grasslands, S.C. Jarvis and B.F. Pain Eds. New York: CAB International.

Agricultural Research Council. 1981. The Nutrient Requirements of Pigs: Technical Review. pp. 307. 2nd ed. Farnham Royal, UK: Commonwealth Agricultural Bureaux.

Alexander, M. 1977. Introduction to Soil Microbiology, 2nd ed. New York: John Wiley & Sons.

American Society of Agricultural Engineers. 1999. Control of manure odors. ASAE EP-379.2. Pp. 655-657. ASAE Standards - 1999, 46th ed. St. Joseph, Michigan.

America's Clean Water Foundation. 2002. On Farm Assessment and Environmental Review (OFAER). http://acwf.org [March 2002].

Amon, M., M. Dobeic, R.W. Sneath, V.R. Phillips, T.H. Misselbrook, and B.F. Pain. 1997. A farm-scale study on the use of clinoptilolite zeolite and De-odorase® for reducing odour and ammonia emissions from broiler houses. Bioresource Technology 61:229-237.

Amon, B., T. Amon, J. Boxberger, and C. Alt. 2001. Emissions of NH_3, N_2O and CH_4 from dairy cows housed in a farmyard manure tying stall. Nutrient Cycling in Agroecosystems 60:103-113.

Andersson, M. 1998. Reducing ammonia emissions by cooling of manure in manure culverts. Nutrient Cycling in Agroecosystems 51:73-79.

Aneja, V.P., J.P. Chauhan, and J.T. Walker. 2000. Characterization of atmospheric ammonia emissions from swine waste storage and treatment lagoons. Journal of Geophysical Research 105:11535-11545.

Aneja, V.P., B. Bunton, J.T. Walker, and B.P. Malik. 2001. Measurement and analysis of atmospheric ammonia emissions from anaerobic lagoons. Atmospheric Environment 35:1949-1958.

Arogo, J., P.W. Westerman, A.J. Heber, W.P. Robarge, and J.J. Classen. 2001. Ammonia Emissions from Animal Feeding Operations. Ames, Iowa: National Center for Manure and Animal Waste Management and Midwest Plan Services.

Arrow, K.J., M.L. Cropper, G.C. Eads, R.W. Hahn, L.B. Lave, R.G. Noll, P.R. Portney, M. Russell, R. Schmalensee, V.K. Smith, and R.N. Stavins. 1996. Is there a role for benefit-cost analysis in environmental, health, and safety regulation? Science 272:221-222.

Auvermann, B., R. Bottcher, A. Heber, D. Meyer, C. B. Parnell, Jr., B. Shaw, and J. Worley. 2001. Particulate Matter Emissions from Confined Animal Feeding Operations: Management and Control Measures. Ames, Iowa: National Center for Manure and Animal Waste Management and Midwest Plan Services.

Barnebey-Cheney. 1987. Scentometer: An Instrument for Field Odor Measurement. Bulletin T-748, Barnebey-Cheney Activated Carbon and Air Purification Equipment Co., Columbus, Ohio. 3 pp.

Barth, C.L., L.F. Elliot, and S.W. Melvin. 1984. Using odor control technology to support animal agriculture. Transactions of the ASAE 27(3):859-864.

Battye, R., W. Battye, C. Overcash, and S. Fudge. 1994. Development and Selection of Ammonia Emission Factors Final Report by EC/R Inc. for U.S. EPA.

Beers, Y. 1957. Introduction to the Theory of Error. Reading, MA: Addison-Wesley Publishing Company. 66 pp.

Berges, M.G.M., and P.J. Crutzen. 1996. Estimates of global N_2O emissions from cattle, pig, and chicken manure, including a discussion of CH_4 emissions. J. Atmos. Chem. 24:241-269.

Boardman, A.E., D.H. Greenberg, A.R. Vining, and D.L. Weimer. 2001. Cost Benefit Analysis, 2nd ed. Upper Saddle River, N.J.: Prentice Hall.

Boehlje, M.D., and V.R. Eidman. 1984. Farm Management. New York: John Wiley & Sons.

Bouwman, A.F. 1996. Direct emission of nitrous oxide from agricultural soils. Nutr. Cycl. Agroecosys. 46:53-70.

Brock, T.D. and M.T. Madigan. 1988. Biology of Microorganisms, 5th ed. Englewood Cliffs, New Jersey: Prentice Hall.

California Environmental Protection Agency. 1998. Environmental Monitoring Branch of the Department of Pesticide Regulation in Sacramento. Available on-line at: http://www.cdpr.ca.gov/docs/pur/vocproj/voc_em.htm [March 2002].

REFERENCES

California Environmental Protection Agency. 1999. Environmental Monitoring Branch of the Department of Pesticide Regulation in Sacramento. Available on-line at: http://www.cdpr.ca.gov/docs/pur/vocproj/voc_em.htm [March 2002].

California Environmental Protection Agency. 2000. Environmental Monitoring Branch of the Department of Pesticide Regulation in Sacramento. Available on-line at: http://www.cdpr.ca.gov/docs/pur/vocproj/voc_em.htm [March 2002].

Chang, C., C.M. Cho, and H.H. Janzen. 1998. Nitrous oxide emission from long-term manured soils. Soil Science Society of America Journal 62:677-682.

Civerolo, K.L., and R.R. Dickerson. 1998. Nitric oxide soil emissions from tilled and untilled cornfields. Ag. Forest Meteor. 90:307-311.

Clayton H., J. Arah, and K.A. Smith. 1994. Measurement of nitrous-oxide emissions from fertilized grassland using closed chambers. J. Geophys. Res.-Atmos. 99:6599-16607.

Cochran, W.G. 1977. Sampling Techniques, 3rd ed. New York:John Wiley & Sons.

Code of Federal Regulations. 2001. Environmental Protection Agency. Subpart E—Alternative Monitoring Systems. Part 75; Sections 40, 41, and 42, pp. 268-273.

Council for Agricultural Science and Technology. 1999. Animal Agriculture and Global Food Supply. II. Series: Task force report No. 135. 92 pp.

Davidson, E.A., and W. Klingerlee. 1997. A global inventory of nitric oxide emissions from soils. Nutrient Cycling Agroecosystems 48:37-50.

Deming, W.E. 1966. Some Theory of Sampling. New York: Dover Publications. 602 pp.

Dendooven, L., L. Duchateau, and J. M. Anderson. 1996. Gaseous products of the denitrification process as affected by the antecedent water regime of the soil. Soil Biology and Biochemistry 28:239-245.

Denmead, O.T. 1997. Progress and challenges in measuring and modelling gaseous nitrogen emissions from grasslands: and overview. Pp 423-438 in Gaseous Nitrogen Emissions from Grasslands, S.C. Jarvis and B.F. Pain, Eds. New York: CAB International.

Dentener, F.J., and P.J. Crutzen. 1994. A 3-dimensional model of the global ammonia cycle. Journal of Atmospheric Chemistry 19:331-369.

Dhandu, A.S. 2001. The non-phytin phosphorus requirement of broilers in the finisher and withdrawal phases of a commercial four phase feeding program. Master of Science Thesis, University of Maryland. College Park, Maryland

Dinn, N.E., J.A. Shelford, and L.J. Fisher. 1998. Use of the Cornell Net Carbohydrate and Protein System and rumen-protected lysine and methionine to reduce nitrogen excretion from lactating dairy cows. J. Dairy

Sci. 81:229-237.

DOE. 2000. Emission of Greenhouse Gases in the United States 2000. Department of Energy Report: DOE/EIA-0573(2000).

Doorn, M.R.J., D.F. Natschke, and P.C. Meeuwissen. 2002. Review of emission factors and methodologies to estimate ammonia emissions from animal waste handling. EPA-600/R-02-017. National Risk Management Research Laboratory, Research Triangle Park, NC 27711.

Dou, Z., R.A. Kohn, J.D. Ferguson, R.C. Boston, and J.D. Newbold. 1996. Managing nitrogen on dairy farms: an integrated approach 1. Model description. J. Dairy Sci. 79:2071-2080.

Dunlap, T.F., R.A. Kohn, G.E. Dahl, M. Varner, and R.A. Erdman. 2000. The impact of somatotropin, milking frequency and photoperiod on dairy farm nutrient flows. J. Dairy Sci. 83:968-976.

Eaton, D. L. 1996. Swine Waste Odor Compounds. Pioneer Hi-Bred International, Livestock Environmental Systems, Des Moines, Iowa. 14 pp.

Ellis, S., J. Webb, T. Misselbrook, and D. Chadwick. 2001. Emissions of ammonia (NH_3), nitrous oxide (N_2O) and methane (CH_4) from a dairy hardstanding in the UK. Nutrient Cycling in Agrosystems 60:115-122.

Environmental Health Sciences Research Center. 2002. Iowa Concentrated Animal Feeding Operations Air Quality Study, Chapter 9. Pp. 184-197. Iowa State University and the University of Iowa Study Group, Final Report, February. Available on-line at http://www.public.health.uiowa.edu/ehsrc/CAFOstudy.htm. [March 2002].

EPA (U.S. Environmental Protection Agency). 1995a. National Air Quality and Emissions Trends Report. Publication Number 454-R-96-005.

EPA (U.S. Environmental Protection Agency). 1995b. Compilation of Air Pollutant Emission Factors AP-42, 5th Ed. Volume I: Stationary Point and Area Sources. Research Triangle Park, N.C.: EPA.

EPA (U.S. Environmental Protection Agency). 1999. U.S. Methane Emissions 1990-2020: Inventories, Projections, and Opportunities for Reductions. Publication Number EPA 430-R-99-013.

EPA (U.S. Environmental Protection Agency). 2001a. Emissions from Animal Feeding Operations (Draft). EPA Contract No. 68-D6-0011.

EPA (U.S. Environmental Protection Agency). 2001b. EPA Air Pollution Control Cost Manual, 6th ed. Publication Number EPA-452-02-001. Research Triangle Park, N.C.: Office of Air Quality Planning and Standards.

EPA (U.S. Environmental Protection Agency). 2002. National Ambient Air Quality Standards. http://www.epa.gov/airs/criteria.html. [March 2002].

Federal Register. 1997. Environmental Protection Agency. National Ambient Air Quality Standards for Particulate Matter 62:38651-38701.

Flessa H, P. Dorsch, and F. Beese. 1995. Seasonal variation of N_2O and CH_4 fluxes in differently managed arable soils in southern germany. Journal of

REFERENCES

Geophysical Research-Atmospheres 100:23115-23124.

Flessa H, P. Dorsch, F. Beese, H. Konig, and A.F. Bouwman. 1996. Influence of cattle wastes on nitrous oxide and methane fluxes in pasture land. Journal of Environmental Quality 25:1366-1370.

Food and Agriculture Organization. 2002. FAO Statistical Databases. Available on-line at: http://apps.fao.org/ [March 2002].

Fritz, J.C., T. Roberts, J.W. Boehne, and E.L. Hove. 1969. Factors affecting the chick's requirement for phosophorus. Poultry Sci. 48:307-320.

Galloway J.N. and E.B. Cowling. 2002. Reactive nitrogen and the world: two hundred years of change. Ambio 31:64-71.

Gibbs, M.J., L. Lewis, and J.S. Hoffman. 1989. Reducing Methane Emissions from Livestock: Opportunities and Issues. Washington, D.C.: U.S. Environmental Protection Agency. 284 pp.

Grelinger, M.A. 1998. Improved Emission Factors for Cattle Feedlots. Emission Inventory: Planning for the Future, Proceedings of Air and Waste Management Association, U.S. Environmental Protection Agency Conference 1:515-524 (October 28-30, 1997).

Grelinger, M.A. and A. Page. 1999. Air pollutant emission factors for swine facilities. Pp. 398-408 in Air and Waste Management Conference Proceedings, October 26-28, 1999.

Groenestein, C.M., and H.G. VanFaassen. 1996. Volatilization of ammonia, nitrous oxide and nitric oxide in deep-litter systems for fattening pigs. J. Agr. Eng. Res. 65:269-274.

Groot Koerkamp P.W.G., J.H.M. Metz, G.H. Uenk, V.R. Phillips, M.R. Holden, R.W. Sneath, J.L. Short, R.P. White, J. Hartung, and J. Seedorf. 1998. Concentrations and emissions of ammonia in livestock buildings in Northern Europe. J. Agr. Eng. Res. 70:79-95.

Grub, W., C.A. Rollo, and J.R. Howes. 1965. Dust Problems in Poultry Environments. Dust and Air Filtration in Animal Shelters (a symposium). American Society of Agricultural Engineers. 338 pp.

Harper, L.A. and R.R. Sharpe. 1998. Pp. 1-22 in Ammonia emissions from swine waste lagoons in the southeastern U.S. coastal plains. Final Report for USDA-ARA Agreement No. 58-6612-7M-022. Division of Air Quality, N.C. Department of Environment and Natural Resources. Raleigh, N.C.

Harper, L.A., R.R. Sharpe, and T.B. Parkin. 2000. Gaseous nitrogen emissions from anaerobic swine lagoons: ammonia, nitrous oxide and dinitrogen gas. Journal Environmental Quality 29:1356-1365.

Heber, A.J., D.S. Bundy, T.T. Lim, J.Q. Ni, B.L. Haymore, C.A. Diehl, and R.K. Duggirala. 1998. Odor emission rates from swine confinement buildings. Pp. 304-310 in Animal Systems and the Environment. Proceedings, International Conference on Odor, Water Quality, Nutrient Management, and Socioeconomic Issues, Des Moines, Iowa. July 20-22.

Heber, A.J., T. Lim, J. Ni, D. Kendall, B. Richert, and A.L. Sutton. 2001. Odor, ammonia and hydrogen sulfide emission factors for grow-finish buildings. (#99-122). Final Report. National Pork Producers Council, Clive, Iowa.

Henry, R.C., C.H. Spiegelman, and S.L Dattner. 1995. Multivariate receptor modeling of Houston Auto-GC VOC Data. Proceedings, APCA annual meeting 10:95.

Hinz, T., and S. Linke, 1998. A comprehensive experimental study of aerial pollutants in and emissions from livestock buildings. Part 2: Results. J. Agr. Eng. Res. 70:119-129.

Hobbs, P.J., T.H. Misselbrook, and T.R. Cumby. 1999. Production and emission of odours and gases from aging pig waste. J. of Agr.l Eng. Res. 72:291-298.

Hoeksma, P., N. Verdoes, J. Ooosthoek, and J.A.M. Voermans. 1982. Reduction of ammonia volatilization from pig houses using aerated slurry as recirculation liquid. Livest. Prod. Sci. 31:121-132.

Holmen, B.A., T.A. James, L.L. Ashbaugh, and R.G. Flocchini. 2001. Lidar-assisted measurement of PM10 emissions from agricultural tilling in California's San Joaquin Valley. Atmospheric Environment 35:3251-3277.

Howarth R. W., E.W. Boyer, W.J. Pabich, and J.N. Galloway. 2002. Nitrogen use in the United States from 1961-2000. Ambio 31:88-98.

Huey, N.A., L.A. Broering, G.A. Jutze, and G.W. Guber. 1960. Objective odor pollution control investigations. Journal of Air Pollution Control Association 10:441

Hutchinson, G.L., A.R. Mosier, and C.E. Andre. 1982. Ammonia and Amine Emissions from a large cattle feedlot. Journal of Environmental Quality 11:288-293.

Intergovernmental Panel on Climate Change. 2001. Climate Change 2001: The Scientific basis. Contributions of Working Group I to the Third Assessment Report of the Intergovernmental Panel on Climate Change. J.T. Houghton, D.J. Griggs, M. Noguer, P.J. van der Linden, X. Dai, K. Maskell, and C.A. Johnson, Eds. Cambridge, U.K. and New York: Cambridge University Press. 881 pp.

Jacobson, L. 1999. Odor and gas emissions from animal manure storage units and buildings. Paper Presented at ASAE Annual Meeting. Toronto, Canada.

James, S.C., and P.R. Eberle. 2000. Economic and Business Principles in Farm Planning and Production. Ames, Iowa: Iowa State University Press.

Jarvis, S.C. 1993. Nitrogen cycling and losses from dairy farms. Soil Use and Management 9:99-105.

Jarvis, S.C. 1997. Emission processes and their interactions in grassland soils. Pp. 1-17 in Gaseous Nitrogen Emissions from Grasslands, S. C. Jarvis and B. F. Pain, Eds. New York: CAB International.

Jarvis S.C., and B.F. Pain. 1994. Greenhouse-gas emissions from intensive livestock systems: their estimation and technologies for reduction. Climatic Change 27:27-38.

REFERENCES

Jiang, J.K., and J.R. Sands. 1998. Report on odour emissions from poultry farms in western australia. Principal Technical Report, Odour Research Laboratory, Centre for Water and Waste Technology, School of Civil and Environmental Engineering, The University of New South Wales. Sydney, Australia.

Johnson, D. E., K.A. Johnson, G.M. Ward and M. E. Branine. 2000. Ruminants and Other Animals. In: Atmospheric Methane: Its Role in the Global Environment, 2nd ed. M.A.K. Khalil (Ed). Springer-Verlag, Berlin. Pp112-133.

Johnson, K.A., and D.E. Johnson. 1995. Methane emissions from cattle. J. Anim. Sci. 73:2483-2492.

Jungbluth, T., E. Hartung, and G. Brose. 2001. Greenhouse gas emissions from animal houses and manure stores. Nutrient Cycling in Agroecosystems 60:122-145.

Kateman, G. and F.W. Pijpers. 1981. Quality Control in Analytical Chemistry. New York: John Wiley & Sons. 276 pp.

Kay, R.D., and W.M. Edwards. 1999. Farm Management. Boston: WCB McGraw Hill.

Kellogg, R.L., C.H. Lander, D.C. Moffitt, and N. Gollehon. 2000. Manure Nutrients Relative to the Capacity of Cropland and Pastureland to Assimilate Nutrients: Spatial and Temporal Trends for the United States. Washington, DC: United States Department of Agriculture.

King, J.J. 1995. The Environmental Dictionary. New York: John Wiley & Sons. 3rd ed.

Kohn, R.A., Z. Dou, J.D. Ferguson, and R.C. Boston. 1997. A sensitivity analysis of nitrogen losses from dairy farms. Journal of Environmental Management 50:417-428.

Kroodsma, W., R. Scholtens, and J. Huis in't Veld. 1988. Ammonia emissions from poultry housing systems volatile emissions from livestock farming and sewage operations. Proceedings of CIGR Seminar Storing, Handling and Spreading of Manure and Municipal Waste 2:7.1-7.13.

Layard, R., and S. Glaister, eds. 1994. Cost Benefit Analysis, 2nd ed. New York: Cambridge University Press.

Leng, R.A. 1993. Quantitative ruminant nutrition—a green science. Australian Journal of Agricultural Research 44:363-380.

Lerner, J., E. Matthews, and I. Fung. 1988. Methane emission from animals: a global high-resolution database. Global Biogeochemical Cycles 2:139-156.

Lessard, R., P. Rochette, E.G. Gregorich, E. Pattey, and R.L. Desjardins. 1996. Nitrous oxide fluxes from manure-amended soil under maize. Journal of Environmental Quality 25:1371-1377.

Leuning, R., S.K. Baker, I.M. Jamie, C.H. Hsu, L. Klein, O.T. Denmead, and D.W.T. Griffith. 1999. Methane emission from free-ranging sheep: a comparison of two measurement methods. Atmospheric Environment.

33:1357-1365.

Li, C.S., S. Frolking, T.A. Frolking. 1992. A model of nitrous oxide evolution from soil driven by rainfall events: I. model structure and sensitivity. J. Geophys. Res. 97:9759-9776.

Li, C.S., V. Narayanan, and R.C. Harriss. 1996. Model estimates of nitrous oxide emissions from agricultural lands in the United States. Global Biogeochemical Cycles 10:297-306.

Lim, T.T., A.J. Heber, J.Q. Ni, A.L. Sutton, and D.T. Kelly. 2001. Characteristics and Emission Rates of Odor from Commercial Swine Nurseries. Transactions of the ASAE. 44:1275–1282.

Lorimor, J., C. Fulhage, R. Zhang, T. Funk, R. Sheffield, D.C. Sheppard, and G.L. Newton. 2001. Manure Management Strategies/Technologies. National Center for Manure and Animal Waste Management and Midwest Plan Services, Ames, Iowa.

MacIntyre, S., R. Wanninkhof, and J.P. Chanton. 1995. Trace gas exchange across the air-water interface in freshwater and coastal marine environments. pp 52-97. in Biogenic Trace Gases: Measuring Emissions from Soil and Water, P.A. Matson and R.C. Harriss, eds. Methods in Ecology. Cambridge Mass: Blackwell Science, Ltd.

McCaughey, W.P., K. Wittenberg, and D, Corrigan. 1999. Impact of pasture type on methane production by lactating beef cows. Canadian Journal of Animal Science. 79:221-226.

Miner, J. R. 1975. Management of odors associated with livestock production. Pp. 378-380 in Managing Livestock Wastes, Proceedings of Third International Symposium on Livestock Wastes. American Society of Agricultural Engineers, St. Joseph, Mich.

Miner, J.R., and R.C. Stroh. 1976. Controlling feedlot surface odor emission rates by application of commercial products. Transactions of the American Society of Agricultural Engineers 19(3):533-538.

Miner, J.R., and L.A. Licht. 1981. Fabric swatches as an aid in livestock manure odor evaluations. Pp. 297-301 in Livestock Wastes: A Renewable Resource. Proceedings of the 4th International Symposium on Livestock Wastes—1980. American Society of Agricultural Engineers, St. Joseph, Mich.

Mosier, A.R., J.M. Duxbury, and J.R. Freney. 1996. Nitrous oxide emissions from agricultural fields: assessment, measurement and mitigation. Plant Soil 181:95-108.

Mosier A, C. Kroeze, C. Nevison, O. Oenema, S. Seitzinger, and O. van Cleemput. 1998. Closing the global N_2O budget: nitrous oxide emissions through the agricultural nitrogen cycle—OECD/IPCC/IEA phase II development of IPCC guidelines for national greenhouse gas inventory methodology. Nutrient Cycling in Agroecosystems 52:225-248.

Muck, R.E. and T.S. Steenhuis. 1982. Nitrogen losses from manure storages.

Agric. Wastes 4:41-48.

Müller, C., R.R. Sherlock, K.C. Cameron, and J.R.F. Barringer. 1997. Application of a mechanistic model to calculate nitrous oxide emissions at a national scale. Pp. 339-349 in Gaseous Nitrogen Emissions from Grasslands, S. C. Jarvis and B. F. Pain, Eds. New York: CAB International.

National Research Council. 1979. Odors from Stationary and Mobile Sources. Washington, DC: National Academy Press.

National Research Council. 1992. Policy Implications of Greenhouse Warming: Mitigation, Adaptation, and Science Base. Washington, DC: National Academy Press.

National Research Council. 1994. Nutrient Requirements of Poultry, 9th revised ed. Washington, DC: National Academy Press.

National Research Council. 1997. Biodiversity II: Understanding and Protecting our Biological Resources. Washington, DC: National Academy Press.

National Research Council. 1998. Nutrient Requirements of Swine, 10th revised ed. Washington, DC: National Academy Press.

National Research Council. 1999. Perspectives on Biodiversity: Valuing Its Role in an Everchanging World. Washington, DC: National Academy Press.

National Research Council. 2000. Nutrient Requirements of Beef Cattle, 7th revised ed. Washington, DC: National Academy Press.

National Research Council. 2001. Nutrient Requirements of Dairy Cattle, 7th revised ed. Washington, DC: National Academy Press.

National Research Council. 2002. The Airliner Cabin Environment and Health of Passengers and Crew. Washington, DC: National Academy Press.

Natrella, M.G. 1963. Experimental Statistics. Statistical Engineering Laboratory, National Bureau of Standards. U.S. Deptartment of Commerce. Washington, D.C.

Ni, J.-Q., A.J. Heber, T.T. Lim, and C.A. Diehl. 2002a. Continuous measurement of hydrogen sulfide emission from two large swine finishing buildings. Journal of Agricultural Science 138:227-236.

Ni, J.-Q., A.J. Heber, C.A. Diehl, T.T. Lim, R.K. Duggirala, and B.L. Haymore. 2002b. Characteristics of hydrogen sulfide concentrations in two mechanically ventilated swine buildings. Canadian Biosystems Engineering 44:11-19.

Ni, J.-Q., A.J. Heber, C.A. Diehl, T.T. Lim, R.K. Duggirala, and B.L. Haymore. 2002c. Summertime concentrations and emissions of hydrogen sulfide at a mechanically-ventilated swine finishing building. Transactions of ASAE 45:193-199.

Ni, J.-Q, A.J. Heber, T.T. Lim, C.A. Diehl, and R.K. Duggirala. 2002d. Hydrogen sulfide from two large pig-finishing buildings with long-term high-frequency measurements. Journal of Agricultural Science 138:227-236.

North Carolina Department of Environmental and Natural Resources – Division of Air Quality. 1999. Status Report on Emissions and Deposition of Atmospheric Nitrogen Compounds from Animal Production in North Carolina.

Oenema, O., A. Bannink, S.G. Sommer, and L. Velthof. 2001. Gaseous nitrogen emissions from livestock farming systems. Pp. 255-289 in Nitrogen in the Environment: Sources, Problems and Management, R.F. Follett, and J.L. Hatfield, Eds. Elsevier. 520 pp.

Ogink, N.W.M., C. van ter Beek, and J.V. Karenbeek. 1997. Odor Emission from Traditional and Low-Emitting Swine Housing Systems: Emission Levels and Their Accuracy. St. Joseph, Mich.: American Society of Agricultural Engineers. ASAE Paper No. 97-4036.

Pace, T.G. 1985. Receptor methods for source apportionment: real world issues and applications. Journal of the Air Pollution Control Association 1149-1153.

Parnell, S.E., B. Lesikar, J.L. Sweeten, and R.E. Lacey. 1994. Determination of an emission factor for cattle feedyards by applying dispersion modeling. Paper 94-4042/94-4082 (Summer 1994): 15 pp.

Paul J.W., and E.G. Beauchamp. 1993. Nitrogen availability for corn in soils amended with urea, cattle slurry, and solid and composted manures. Can. J. Soil Sci. 73:253-266.

Penner, J.E., M. Andreae, A. Annegarn, L. Barrie, J. Feichter, D. Hegg, A. Jayaraman, R. Leaitch, D. Murphy, J. Nganga, and G. Pitari. 2001. Aerosols, their direct and indirect effects. In Climate Change 2001: The Scientific Basis. Contributions of Working Group I to the Third Assessment Report of the Intergovernmental Panel on Climate Change. J.T. Houghton, D.J. Griggs, M. Noguer, P.J. van der Linden, X. Dai, K. Maskell, C.A. Johnson, Eds. Cambridge, U.K. and New York: Cambridge University Press. 881 pp.

Peters, J.A. and T.R. Blackwood. 1977. Source Assessment: Beef Cattle Feedlots. EPA-600/ 2-77-107 U.S. Environmental Protection Agency, Research Triangle Park, N.C. 101 pp.

Petersen, S.O. 1999. Nitrous oxide emissions from manure and inorganic fertilizers applied to spring barley. Journal of Environmental Quality 28:1610-1618.

Powers, W.J., H.H. Van Horn, A.C. Wilkie, C.J. Wilcox, and R.A. Nordstedt. 1999. Effects of anaerobic digestion and additives to effluent or cattle feed on odor and odorant concentration. J. Anim. Sci. 77:1412-1421.

Prather M., D. Ehalt, F. Dentener, R. Derwent, Dlugokencky, E. Holland, I. Isaksen, J. Katima, Kirchhoff, P. Matson, P. Midgley, and M. Wang. 2001. Atmospheric chemistry and greenhouse gases. In Climate Change 2001:

The Scientific Basis. Contributions of Working Group I to the Third Assessment Report of the Intergovernmental Panel on Climate. J.T. Houghton, D.J. Griggs, M. Noguer, P.J. van der Linden, X. Dai, K. Maskell, and C.A. Johnson, Eds. Cambridge, U.K. and New York: Cambridge University Press. 881 pp.

Purdue, L.J. 1992. Determination of the strong acidity of atmospheric fine-particles using annular denuder technology. Enhanced method. EPA/600/R-93/037 (NTIS PB93-178234).

Ramadan, Z., X.-H. Song, P.K. Hopke. 2000. Identification of sources of Phoenix aerosol by positive matrix factorization. J. Air Waste Manage. Assoc. 50:1308-1320.

Robarge, W.P., J.T. Walker, R.B. McCulloch, and G. Murray. 2002. Atmospheric concentrations of ammonia and ammonium at an agricultural site in the Southeast United States. Atmospheric Environment 36:1661-1674.

Robertson G.P., E.A. Paul, and R.R. Harwood. 2000. Greenhouse gases in intensive agriculture: Contributions of individual gases to the radiative forcing of the atmosphere. Science 289:1922-1925.

Rochette, P., E. van Bochove, D. Prevost, D.A. Angers, D. Cote, and N. Bertrand. 2000. Soil carbon and nitrogen dynamics following application of pig slurry for the 19th consecutive year: II. Nitrous oxide fluxes and mineral nitrogen. Soil Science Society of America Journal 64:1396-1403.

SCAQMD (South Coast Air Quality Management District). 1993. Projected Air Quality as a Result of Reducing Emissions from the Livestock Industry in the South Coast Air Basin.

Schiffman, S.S. and C.M. Williams. 1999. Evaluation of swine odor control products using human odor panels. Pp 110-118 in Animal Waste Management Symposium. Raleigh, NC State University.

Schiffman, S.S., E.A. Sattely Miller, M.S. Suggs, and B.G. Graham. 1995. The effect of environmental odors emanating from commercial swine operations on the mood of nearby residents. Brain Research Bulletin 37:369-375.

Schiffman, S.S., J.L. Bennett, and J.H. Raymer. 2001. Quantification of odors and odorants from swine operations in North Carolina. Agricultural Forest Meteorology. 108:213-240.

Schmidt, D. 2000. Odor, hydrogen sulfide, and ammonia emissions from composting of caged layer manure. Final report to Broiler and Egg Association of Minnesota.

Skiba U., D. Fowler, and K.A. Smith. 1997. Nitric oxide emissions from agricultural soils in temperate and tropical climates: sources, controls and mitigation options. Nutr.Cycl. Agroecosys. 48:139-153.

Slemr, F. and W. Seiler. 1984. Field emissions of NO and NO_2 from fertilized and unfertilized soils. J. Atmos. Chem. 2:1-24.

Smith K.A., I.P. McTaggart, and H. Tsuruta. 1997. Emissions of N_2O and NO associated with nitrogen fertilization in intensive agriculture, and the potential for mitigation. Soil Use Manage. 13:296-304.

St-Pierre, N.R. and C. S. Thraen. 1999. Animal grouping strategies, sources of variation, and economic factors affecting nutrient balance on dairy farms. J. Anim. Sci. Suppl. 2 77:73-83.

Sutton, A.L., K.B. Kephart, J.A. Patterson, R. Mumma, D.T. Kelly, E. Bogus, D.D. Jones, and A.J. Heber. 1996. Manipulating swine diets to reduce ammonia and odor emissions. Pp. 445-452 in International Conference on Air Pollution from Agricultural Operations, Kansas City, M.O., February 7-9.

Sweeten, J.M., D.L. Reddell, L. Schake, and B. Garner. 1977. Odor Intensities at Cattle Feedlots. Transactions of the American Society of Agricultural Engineers 20(3):502-508.

Sweeten, J.M., D.L. Reddell, A.R. McFarland, R.O. Gauntt, and J.E. Sorel. 1983. Field measurement of ambient odors with a butanol olfactometer. Transactions of the American Society of Agricultural Engineers 26(4):1206-1216.

Sweeten, J.M., R.C. Childers, J.S. Cochran, and R. Bowler. 1991. Odor control from poultry manure composting plant using a soil filter. Applied Engineering in Agriculture 7(4):439-449.

Sweeten, J.M., C.B. Parnell, B.W. Shaw, B.W. Auvermann. 1998. Particle size distributions of cattle feedlot dust emissions. Transactions of ASAE 41(5):1477-148. St. Joseph, Mich.

Sweeten, J.M, L.D. Jocobson, A.J. Heber, D.R. Schmidt, J.C. Lorimor, P.W. Westerman, J.R. Miner, R. Zhang, C.M. Williams, and B.W. Auverman. 2001. Chapter 1 in Odor Mitigation for Concentrated Animal Feeding Operations. [CD ROM] National Center for Manure and Animal Waste Management (USDA), Ames, Iowa.

Takai, H., S. Pedersen, J.O. Johnsen, J.H.M. Metz, P.W.G. Groot Koerkamp, G.H. Uenk, V.R. Phillips, M.R. Holden, R.W. Sneath, J.L. Short, R.P. White, J. Hurtung, J. Seedorf, M. Schroder, K.H. Linkert, and C.M. Wathes. 1998. Concentrations and emissions of airborne dust in livestock buildings in Northern Europe. Journal of Agricultural Engineering Research 70:59-70.

Tate, R.L. 1995. Soil Microbiology. New York: John Wiley & Sons.

Taylor, J.K. 1987. Quality Assurance of Chemical Measurements. Chelsa, Mich.: Lewis Publishers. 328 pp.

Thompson, R.B., J.C. Ryden, and D.R. Lockyer. 1987. Fate of nitrogen in cattle slurry following surface application or injection to grassland. J. Soil Sci. 38:689-700.

Thu, K., K. Donham, R. Ziegenhorn, S.J. Reynolds, P.S. Thorne, P. Subramanian, W. Whitten, and J. Stookesberry. 1997. A control study of health and quality of life of residents living in the vicinity of large scale

swine production facilities. J. Agric. Health Safety 3:13-26.
Todd, L.A., M. Ramanathan, K. Mottus, R. Katz, A. Dodson, and G. Mihan. 2001. Measuring chemical emissions using open-path Fourier transform infrared (OP-FTIR) spectroscopy and computer-assisted tomography. Atmospheric Environment 35:1937-1947.
USDA (U.S. Department of Agriculture). 2000. Confined Livestock Air Quality Subcommittee, J.M. Sweeten, Chair. Air Quality Research and Technology Transfer Programs for Concentrated Animal Feeding Operations. USDA Agricultural Air Quality Task Force (AAQTF) Meeting, Washington, DC.
USDA (U.S. Department of Agriculture). 2002. Agricultural Research Service Air Quality National Program. Available on-line at http://www.nps.ars.usda.gov/programs/programs.htm?NPNUMBER=203 [2002, March].
van Aardenne J.A., F.J. Dentener, C.G.M. Klijn Goldewijk, J. Lelieveld, and J.G.J. Olivier. 2001. A $1°-1°$ resolution dataset of historical anthropogenic trace gas emissions for the period 1890-1990. Global Biogeochem. Cycles. 15:909.
Verdoes, N. and N.W.M. Ogink. 1997. Odour emission from pig houses with low ammonia emission. Pp. 317-325 in Proceedings of the International symposium on Ammonia and Odour Control from Animal Production Facilities, Vinkeloord, The Netherlands. J.A.M. Voermans, and G. Moteny, Eds.
Wagner-Riddle, C., G.W. Thurtell, G.K. Kidd, E.G. Beauchamp, and R. Sweetman. 1997. Estimates of nitrous oxide emissions from agricultural fields over 28 months. Canadian Journal of Soil Science 77:135-144.
Wathes, C.M., V.R. Phillips, M.R. Holden, R.W. Sneath, J.L. Short, R.P. White, J. Hartung, J. Seedorf, M. Schroder, K.H. Linkert, S. Pedersen, H. Takai, J.O. Johnsen, P.W.G. Groot Koerkamp, G.H. Uenk, J.H.M. Metz, T. Hinz, V. Caspary, and S. Linke. 1998. Emissions of aerial pollutants in livestock buildings in Northern Europe: overview of a multinational project. J. Agric. Engng. Res. 70:3-9.
Watson, J.G., N.F. Robinson, C. Lewis, and T. Coulter. 1997. Chemical mass balance receptor model version 8 (CMB8) user's manual. Desert Research Institute, Document No. 1808.1D1, Reno, Nev.
Watson, J.G., J.C. Chow, and E.M. Fujita. 2001. Review of volatile organic compound source apportionment by chemical mass balance. Atmos. Environ. 35:1567-1584.
Watts, P.J. 1991. The Measurement of Odours: A Discussion Paper for Australia. AMLRDC Report No. DAQ. 64/5. Feedlot Services Group, Queensland Department of Primary Industries, Toowoomba. 40 Pp.
Whalen S.C., R.L. Phillips, and E.N. Fischer. 2000. Nitrous oxide emission from an agricultural field fertilized with liquid lagoonal swine effluent. Global

Biogeochemical Cycles 14:545-558.

Williams, C.M., and S.S. Schiffman. 1996. Effect of liquid swine manure additives on odor parameters. Pp. 409-412 in International Conference on Air Pollution from Agricultural Operations. Kansas City, Mo.: American Society of Agricultural Engineers.

Williams, E.J., A. Guenthre, and F.C. Fehsenfeld. 1992. An inventory of nitric oxide emissions from soils in the United States. J. Geophys. Res. 97:7511-7519.

Williams, D.L., P. Ineson, and P.A. Coward. 1999. Temporal variations in nitrous oxide fluxes from urine-affected grassland. Soil Biology and Biochemistry 31:779-788.

Williams, C.M. July 24, 2001. Technology Determinations per Agreements Between Attorney General of North Carolina and Smithfield Foods, Inc. and Premium Standard Farms Year One Progress Report. Available on-line at http://www.cals.ncsu.edu/waste_mgt/progress.pdf. [2002, March]

Wing, S., and S. Wolf. 2000. Intensive livestock operations, health, and quality of life among eastern North Carolina residents. Environmental Health Persp. 108:233-238.

Wood, S.L., K.A. Janni, C.J. Clanton, D.R. Schmidt, L.D. Jacobson, and S. Weisberg. 2001. Odor and air emissions from animal production systems. Paper presented at ASAE Annual International Meeting. Paper No. 014043.

Yamulki, S., S.C. Jarvis, and P. Owen. 1998. Nitrous oxide emissions from excreta applied in a simulated grazing pattern. Soil Biology and Biochemistry 30:491-500.

Yienger, J.J., and H. Levy II. 1995. Empirical model of soil biogenic NO_x emissions. J. Geophys. Res. 100:11447-11464.

Zahn, J.A., J.L. Hatfield, Y.S. Do, A.A. DiSpirito, D.A. Laird, and R.L. Pfeiffer. 1997. Characterization of volatile organic emissions and wastes from a swine production facility. J. Environ. Qual. 26:1687-1696.

Zahn, J.A., A.E. Tung, B.A. Roberts and J.L. Hatfield. 2001. Abatement of ammonia and hydrogen sulfide emissions from a swine lagoon using a polymer biocover. Journal of the Air and Waste Management Association 51:562-573.

Zhu, J., L.D. Jacobson, D.R. Schmidt, and R.E. Nicolai. 1999. Daily Variation in Odor and Gas Emissions from Animal Facilities. ASAE Paper No. 99-4146. St. Joseph, Mich.: American Society of Agricultural Engineers.-

Zhu, J., L. Jacobson, D. Schmidt and R. Nicolai. 2000. Daily variations in odor and gas emissions from animal facilities. American Society of Agricultural Engineers 16:153-158.

Appendix A
Statement of Task

An ad hoc committee of the standing Committee on Animal Nutrition will be appointed to conduct a rigorous scientific review of air emission factors as related to current animal feeding and production systems in the United States. The committee will review and evaluate the scientific basis for estimating the emissions of various air pollutants (PM, PM10, PM2.5, hydrogen sulfide, ammonia, odor, VOCs, methane, and nitrous oxide) from confined livestock and poultry production systems to the atmosphere. In its evaluation, the committee will review characteristics of agricultural animal industries, methods for measuring and estimating emissions, and potential best management practices, including costs and technologic feasibility. The committee will focus on confined animal feeding production systems and will evaluate them in terms of biologic systems. The committee will consider all relevant literature and data, including reports compiled by the EPA and USDA on air quality research, air emissions, and air quality impacts of livestock waste. The study will identify critical short- and long-term research needs and will provide recommendations on the most promising science-based methodologic and modeling approaches for estimating and measuring emissions—including deposition, rate, cycle, fate, and transport—as well as on potential mitigation technologies. The committee will issue an interim report including a review of methodologies and data presented in " Air Emissions From Animal Feeding Operations" EPA Office of Air and Radiation, August 15, 2001.

Appendix B

Glossary

Accuracy: The closeness of an individual measurement or of the average of a number of measurements to the true value. Deviation from the true value is a measure of bias in the individual measurement or averaged value.
ACWF: America's Clean Water Foundation
AER: Allowable emission rate
AFO: Animal feeding operation
Animal feeding operation: As defined by the U.S. Environmental Protection Agency (40 CFR 122.23): a "lot or facility" where animals "have been, are, or will be stabled or confined and fed or maintained for a total of 45 days or more in any 12 month period and crops, vegetation, forage growth, or post-harvest residues are not sustained in the normal growing season over any portion of the lot or facility."
Animal unit: A unit of measure that is used to compare different animal species:
    1. EPA (66 FR 2960- 3138): 1 cattle excluding mature dairy and veal cattle; 0.7 mature dairy cattle; 2.5 swine weighing over 55 pounds; 10 swine weighing 55 pounds or less; 55 turkeys; 100 chickens; and 1 veal calf.
    2. USDA: 1,000 pounds of live animal weight
Anthropogenic: Caused by humans
ARS: Agricultural Research Service (USDA)
AU: Animal unit
BW: Body weight
C: Carbon

APPENDIX B

C₃, C₄, etc: Molecules with 3, 4, etc. carbon atoms
CCN: Cloud condensation nuclei
CH$_4$: Methane
CNMP: Comprehensive nutrient management plans
CO$_2$ equivalent: The mass of CO$_2$ with the same climate change potential as the mass of the greenhouse gas in question
d: Day(s)
Denitrification: Reduction of nitrates or nitrites to nitrogen-containing gases (mostly N$_2$)
EPA: U.S. Environmental Protection Agency
FID: Flame-ionization detector
GC: Gas chromatography
ha: hectare; an area 100 meters square, or about 2.5 acres
H$_2$S: Hydrogen sulfide
IPCC: Intergovernmental Panel on Climate Change
kg: kilogram, or 1,000 grams (about 2.2 pounds)
km: kilometer, or 1,000 meters
Lidar: A device similar to radar except that it emits pulsed laser light rather than microwaves
LU: Live unit, 500 kg of body weight
Manure: A mixture of animal feces and urine, that may also include litter or bedding materials
MS: Mass spectrometer
MT: Million tones
µm: micrometer (10^{-6}m); micron
N: Nitrogen
N$_2$: Dinitrogen molecule
NAAQS: National Ambient Air Quality Standards
NH$_3$: Ammonia
Nitrification: Oxidation of an ammonia compound to nitric acid, nitrous acid, or any nitrate or nitrite, especially by the action of nitrobacteria
nm: nanometer; 10^{-9}m
NO: Nitric oxide
NO$_x$: Nitric oxide and nitrogen dioxide rapidly interconverted in the atmosphere
NO$_y$: The sum of all oxidized nitrogen species in the atmosphere
N$_2$O: Nitrous oxide
NRC: National Research Council
NRCS: National Resource Conservation Service
Nutrient excretion factor: an estimate of an element, for example nitrogen, excreted by an animal usually reported as kg per day (or year) per animal (animal unit or kg of bodyweight).
OFAER: On Farm Assessment and Environmental Review (of the ACWF)
OH: Hydroxy radical

Orographic: Relating to the physical geography of mountains and mountain ranges
PAN: Peroxyacetyl nitrate
PBL: Planetary boundary layer
PM: Particulate matter
PM2.5: Particulate matter having an aerodynamic diameter of 2.5 µm or less
PM10: Particulate matter having an aerodynamic diameter of 10 µm or less
ppb: Parts per billion by volume
ppm: Parts per million by volume
Precision: Agreement among individual measurements of the same property, under prescribed similar conditions
S: Sulfur
SIP: State implementation plans for NAAQS
Sulfuric acid: H_2SO_4
Synoptic: Of or relating to data obtained nearly simultaneously over a large area of the atmosphere
Tg: Teragram, 1×10^{12} g
TSP: Total suspended particulates
Uncertainty: The degree of confidence that can be assigned to a numerical measurement in terms of both its accuracy and its precision
USDA: U.S. Department of Agriculture
VOC: Volatile organic compound
Volatile solids: Weight lost upon ignition at 550 °C (using Method 2540 E of the American Public Health Association). Volatile solids provide an approximation of moisture and organic matter present.
yr: year(s)

Appendix C
Public Meeting Agendas

January 7, 2002 – Washington D.C.

1:00	**Sponsor Perspective, EPA**
	Randy Waite, USEPA-OAR
	Renee Johnson, USEPA-OW
1:30	**Issues at the Interface of Animal Agriculture and Air Quality**

    Technical Assistance Perspectives
      Thomas Christensen, Director
      USDA-NRCS Animal Husbandry and Clean Water Programs Division
    Societal and Environmental Considerations
      Dr. Joseph Rudek, Senior Scientist
      Environmental Defense
    Industry Approaches and Dynamics
      David Townsend, Vice President of Environmental Affairs
      Premium Standard Farms Research and Development

3:15-3:30	**Break**
3:30	**Comments from Participants Registered to Present**
4:15	**Input from Other Participants**

January 24, 2002 – Raleigh, North Carolina

7:00 PM   Roundtable Discussion with "Air Emissions From Animal Feeding Operations" Report Authors (August 15, 2001 Draft. EPA Contract No. 68-D6-0011 Task Order 71.)
John H. Martin Jr, Hall Associates
Roy V. Oommen, Eastern Research Group
John D. Crenshaw, Eastern Research Group

8:30 PM   Adjourn

January 25, 2002 – Raleigh, North Carolina

8:00 AM   Introduction
Perry Hagenstein, Chair
NRC Committee on Air Emissions from Animal Feeding Operations

8:10   **In-ground Digestor with Biogas Recovery and Electricity Generation**
Dr. Leonard Bull, Associate Director Animal and Poultry Waste Center
North Carolina State University
Raleigh, NC

8:30   **Measurement of Trace-Gas Emissions In Animal Production Systems**
Dr. Lowry Harper, Research Scientist
United States Department of Agriculture
Watkinsville, GA

8:50   **Open Path Laser Technology/Modeling to Derive Emission Factors for Swine Production Facilities**
Dr. Bruce Harris, Research Scientist
Environmental Protection Agency
Research Triangle Park, NC

9:10   **Pathogens and Air Quality Concerns**
Dr. Mark Sobsey, Professor Environmental Sciences and Engineering
University of North Carolina
Chapel Hill, NC

9:30   **Questions**
Robert Flocchini, Vice-Chair
NRC Committee on Air Emissions from Animal Feeding Operations

APPENDIX C   95

9:45	**Break**
10:00	**Permeable Lagoon Cover for Odor and Ammonia Volatilzation Reduction**
	Dr. Leonard Bull, Associate Director Animal and Poultry Waste Center
	North Carolina State University
	Raleigh, NC
10:20	**Odor Quantification and Environmental Concerns**
	Dr. Susan Schiffman, Professor of Medical Psychology
	Duke University
	Durham, NC
10:40	**Technology for Mitigating PM and Odors from Buildings**
	Dr. Bob Bottcher, Professor of Biological and Agricultural Engineering
	North Carolina State University
	Raleigh, NC
11:00	**Annual Denuder Technology**
	John T. Walker, Chemist
	Environmental Protection Agency
	Research Triangle Park, NC
11:20	**Additional Questions**
	Robert Flocchini
11:30	**Sponsor Perspective**
	Sally Shaver
	Division Director Office of Air Quality Planning and Standards
	Environmental Protection Agency
	Research Triangle Park, NC
11:50	**General Discussion**
	Perry Hagenstein
12:00 PM	**Adjourn**

February 24, 2002 – Denver, Colorado

Monitoring Air Emissions Through Microclimate Meteorological Techniques

1:30	**Introduction**
	Perry Hagenstein, Chair
	NRC Committee on Air Emissions from Animal Feeding Operations

1:40	**Surface Exchange Flux Measurements Utilizing the National Center for Atmospheric Research Integrated Surface Flux Facility** *Dr. Tony Delany, Engineer IV* *Atmospheric Technology Division* *National Center for Atmospheric Research* *Boulder, CO*
2:00	**Flux Footprint Considerations for Micrometeorological Flux Measurement Techniques** *Dr. Tom Horst* *Atmospheric Technology Division* *National Center for Atmospheric Research* *Boulder, CO*
2:20	**Micrometeorological Methods for Estimating VOC and Ammonia fluxes** *Dr. Alex Guenther, Scientist II* *Atmospheric Chemistry Division* *National Center for Atmospheric Research* *Boulder, CO*
2:40	**Analysis of Single Aerosol Particles with a Mass Spectrometer** *Dr. Daniel Murphy* *Aeronomy Laboratory* *National Oceanographic and Atmospheric Administration* *Boulder, CO*
3:00	**Questions and General Discussion** *Robert Flocchini, Vice-Chair* *NRC Committee on Air Emissions from Animal Feeding Operations*
3:15	**Break**

Air Emission Measurement and Mitigation for Beef Feedlots

3:30	**Introduction** *Perry Hagenstein, Chair*
3:40	**Odor Measurement and Mitigation** *Dr. John Sweeten, Professor and Resident Director* *Agricultural Research & Extension Center* *Texas A&M University* *Amarillo, TX*
4:00	**Methane Production from Livestock and Mitigation** *Dr. Don Johnson, Professor* *Department of Animal Sciences* *Colorado State University*

	Fort Collins, CO
4:20	**Mitigation Technology**
	Dr. Bob McGregor
	Water and Waste
	Denver, CO
4:40	**Questions and General Discussion**
	Robert Flocchini, Vice-Chair
5:00	**Comments from Participants Registered to Present**
5:30	**Input from Other Participants**

Appendix D
Twenty-Three Model Farms Described By EPA

		Elements of Model Farms			
Animal	Model Farm ID	Confinement and Manure Collection System	Solids Separation Activities	Manure Storage and/or Stabilization	Land Application
Beef	B1A	Drylot (scraped)	Solids separation for runoff (using a settling basin)	Storage pond and stockpile	Liquid and solid
Beef	B1B	Drylot (scraped)	No solids separation	Storage pond and stockpile	Liquid and solid
Dairy	D1A	Freestall barn (flush); milking center (flush); drylot (scraped)	Solids separation	Anaerobic lagoon and stockpile	Liquid and solid
Dairy	D1B	Freestall barn (flush); milking center (flush); drylot (scraped)	No solids separation	Anaerobic lagoon and stockpile	Liquid and solid

APPENDIX D

Animal	Model Farm ID	Confinement and Manure Collection System	Solids Separation Activities	Manure Storage and/or Stabilization	Land Application
		Elements of Model Farms (continued)			
Dairy	D2A	Freestall barn (scrape); milking center (flush); drylot (scraped)	Solids separation	Anaerobic lagoon and stockpile	Liquid and solid
Dairy	D2B	Freestall barn (scrape); milking center (flush); drylot (scraped)	No solids separation	Anaerobic lagoon and stockpile	Liquid and solid
Dairy	D3A	Milking center (flush); drylot (scraped)	Solids separation	Storage pond and stockpile	Liquid and solid
Dairy	D3B	Milking center (flush); drylot (scraped)	No solids separation	Storage pond and stockpile	Liquid and solid
Dairy	D4A	Drylot feed alley (flush); milking center (flush); drylot (scraped)	Solids separation	Storage pond and stockpile	Liquid and solid
Dairy	D4B	Drylot feed alley (flush); milking center (flush); drylot (scraped)	No solids separation	Storage pond and stockpile	Liquid and solid
Poultry-broilers	C1A	Broiler house w/bedding	None	Covered cake and open litter	Solid
Poultry-broilers	C1B	Broiler house w/bedding	None	Covered cake	Solid
Poultry-layers	C2	Caged layer high rise house	None	None	Solid

Animal	Model Farm ID	Confinement and Manure Collection System	Solids Separation Activities	Manure Storage and/or Stabilization	Land Application
Poultry-layers	C3	Caged layer house (flush)	None	Anaerobic lagoon	Liquid
Poultry-turkey	T1A	Turkey house w/bedding	None	Covered cake and open litter	Solid
Poultry-turkey	T1B	Turkey house w/bedding	None	Covered storage of cake	Solid
Swine	S1	Enclosed house (flush)	None	Anaerobic lagoon	Liquid
Swine	S2	Enclosed house (pit recharge)	None	Anaerobic lagoon	Liquid
Swine	S3A	Enclosed house (pull plug pit)	None	Anaerobic lagoon	Liquid
Swine	S3B	Enclosed house (pull plug pit)	None	External tank or pond	Liquid
Swine	S4	Enclosed house w/pit	None	None	Liquid
Veal	V1	Enclosed house (flush)	None	Anaerobic lagoon	Liquid
Veal	V2	Enclosed house w/pit	None	None	Liquid

SOURCE: Adapted from EPA (2001a, Table 1).

Appendix E
About the Authors

Perry R. Hagenstein, Ph.D. (Chair) is president of the Institute for Forest Analysis, Planning, and Policy, a non-profit research and education organization. Prior to this, he was executive director of the New England Natural Resources Center and served as a Charles Bullard Research Fellow at the John F. Kennedy School of Government at Harvard. He also served as senior policy analyst for the U.S. Public Land Law Review Commission and was a principal economist for the USDA Forest Service. Hagenstein received his B.S. (1952) from the University of Minnesota, M.F. (1953) from Yale University, and Ph.D. (1963) in forest and natural resources economics from the University of Michigan. He currently serves on the NRC Board on Agriculture and Natural Resources and previously on the Board on Earth Sciences and Resources and Board on Mineral and Energy Resources. Hagenstein has served on nine prior NRC committees including the Committee on Noneconomic and Economic Value of Biodiversity: Application for Ecosystem Management, Committee on Hardrock Mining on Federal Lands (Chair), the Committee on Onshore Oil and Gas Leasing (Chair), and the Committee on Abandoned Mine Lands (Chair).

Robert G. Flocchini, Ph.D. (Vice-Chair) is professor of the Department of Land, Air and Water Resources and director of the Crocker Nuclear Laboratory at University of California, Davis. His interests include the identification, transport, and fate of particulate matter with regard to agricultural sources and application of nuclear techniques for emission measurement and characterization in agriculture and environment. He received his B.A. (1969) from the University of San Francisco, and M.A. (1971) and Ph.D. (1974) in Physics from the University of California, Davis. Flocchini currently serves as a member of the USDA Task Force on Agricultural Air Quality and trustee of the National Institute for Global Environmental Change.

John C. Bailar III, M.D., Ph.D. is Professor Emeritus at the University of Chicago. He is a retired commissioned officer of the U.S. Public Health Service, and worked for the National Cancer Institute for 22 years. He has also held academic appointments at Harvard University and McGill University. Dr. Bailar's research interests include assessing health risks from chemical hazards and air pollutants and interpreting statistical evidence in medicine, with a special emphasis on cancer. Bailar received his B.A. (1953) from the University of Colorado, M.D. (1955) from Yale University, and Ph.D. (1971) in statistics from American University. He is a member of the Institute of Medicine and has served on over twenty NRC committees including the Committee on Estimating the Health-Risk-Reduction Benefits of Proposed Air Regulations (Chair), Committee on Risk Assessment of Hazardous Air Pollutants, and Committee on Epidemiology of Air Pollutants.

Candis Claiborn, Ph.D., is an associate professor in the Department of Civil and Environmental Engineering at Washington State University. Prior to that she was a senior process control engineer at ARCO Petroleum Products and a process engineer at Chevron. Her areas of expertise include airborne particulate matter measurement, characterization, and emissions, and air pollution control. She received her B.S. (1980) in chemical engineering from the University of Idaho and Ph.D. (1991) from North Carolina State University. Claiborn was a member of the Western Governor's Association's "Western Regional Air Partnership Expert Panel on Windblown and Mechanically Generated Fugitive Dust" and contributing author for the USEPA "Air Quality Criteria Development for Particulate Matter".

Russell R. Dickerson, Ph.D. is professor and chair (effective 1 July 2002) of the Department of Meteorology at the University of Maryland, College Park. Prior to Maryland, he worked at the National Center for Atmospheric Research and at the Max Planck Institute for Chemistry, in Mainz, Germany. He received his A.B. (1975) from the University of Chicago, and M.S. (1978) from the University of Michigan, and Ph.D. (1980) in Chemistry from the University of Michigan. His areas of expertise include atmospheric chemistry, air pollution, and biogeochemical cycles with an emphasis on NOx, O_3, CO, black carbon, and ammonia. Dickerson previously served on the NRC Panel to Review the

Langley Distributed Active Archive Center (DAAC) and US/Mid East Research Grants Panel.

James N. Galloway, Ph.D. is professor Department of Environmental Sciences at the University of Virginia and is currently a visiting scientist at the Marine Biological Laboratory and the Woods Hole Oceanographic Institution. His major interests include the biogeochemistry of emissions, transport, and fate of nitrogen and sulfur and their potential effects on ecology. He received his B.A. (1966) from Whittier College and Ph.D. (1972) in chemistry from the University of California, San Diego. Galloway has given expert testimony to state and federal agencies and legislatures on environmental issues. Galloway has previously served on the NRC Global Climate Change Study Panel (Chair), Panel on Processes of Lake Acidification, Tri-Academy Committee on Acid Deposition, and Committee on Transport and Transformation Chemistry in Acid Deposition.

Margaret Rosso Grossman, Ph.D., J.D., is professor of agricultural law in the Department of Agricultural and Consumer Economics at the University of Illinois. She has spent sabbatical leaves (1986-87, 1993-94, 2000-01) and many summers in the Law and Governance Group (formerly Department of Agrarian Law) at Wageningen University, The Netherlands. Her research interests include domestic and international agricultural and environmental law. She received her B. Mus. (1969) from the University of Illinois, A.M. (1970) from Stanford University, Ph.D. (1977) from the University of Illinois, and J.D. (1979) from the University of Illinois. Grossman is past president (1991) of the American Agricultural Law Association and received the AALA Distinguished Service Award (1993). She was awarded the Silver Medal of the European Council for Agricultural Law (1999), and she has received three Fulbright grants to support her research in Europe. Grossman is a member of the Bar in Illinois and the District of Columbia (inactive).

Prasad Kasibhatla, Ph.D. is associate professor Division of Environmental Science and Policy at Duke University. His areas of expertise include: tropospheric chemistry and transport; global tropospheric oxidants; global tropospheric aerosols; regional air quality; anthropogenic impacts on atmospheric composition and ecosystems; and global and regional tropospheric chemistry modeling. He received his B.S. (1982) from the University of Bombay, M.S. (1984) from the University of Kentucky, and Ph.D. (1988) in chemical engineering from the University of Kentucky. Kasibhatla has previously served on the NASA Committee for Measurement of Air Pollution from Satellites and review panels for NOAA and DOE atmospheric chemistry programs.

Richard A. Kohn, Ph.D. is associate professor Department of Animal and Avian Sciences at the University of Maryland. His areas of expertise include environmental impact of animal production systems; effect of diet on nitrogen and phosphorous excretion; and modeling of nutrient metabolism and whole farm nutrient management. He received his B.S. (1985) from Cornell University, M.S. (1987) from the University of New Hampshire, and Ph.D. (1993) in animal science from Michigan State University all in animal science. In 1999, Kohn gave an invited presentation on "Calculating the environmental impact of animal feeding and management" to the NRC Committee on Animal Nutrition.

Michael P. Lacy, Ph.D. is professor and chair Department of Poultry Science at the University of Georgia. His area of expertise is poultry, specifically, production and management; housing and equipment; ventilation; management in hot climates; and mechanical harvesting. Lacy received his B.S. (1974), M.S. (1982), and Ph.D. (1985) from the Virginia Polytechnic Institute & State University.

Calvin B. Parnell, Jr., Ph.D., PE is a Regents professor of the Department of Biological and Agricultural Engineering (BAEN) at Texas A&M University. He has special expertise in the air pollution regulatory process, including permitting and enforcement of air pollution regulations. His research expertise includes pollutant measurements, dispersion modeling, emission factor development, and air pollution abatement. In addition, Parnell is known for his expertise of agricultural processing, grain dust explosions, and energy conversion of biomass. He received his B.S. (1964) from New Mexico State University, M.S. (1965) from Clemson University, and Ph.D. (1970) in environmental systems engineering from Clemson University. Parnell is a registered professional engineer in Texas, Fellow of the American Society of Agricultural Engineers, and a member of the Air and Waste Management Association. He has provided expert testimony to state and federal legislatures on agricultural air quality. Parnell has previously served on the Texas Air Control Board and currently serves on the USDA Task Force on Agricultural Air Quality. He currently receives research funding from a Texas Legislative Initiative on "Air Pollution Regulatory Impacts on Agricultural Operations". Parnell teaches undergraduate and graduate courses in air pollution engineering.

Robbi Pritchard, Ph.D. is professor Department of Animal and Range Sciences at South Dakota State University. His interests include beef feedlot management and ruminant nutrition. Pritchard received his A.A. (1975) from Black Hawk Junior College, B.S. (1977) and M.S. (1978) from Southern Illinois University, and Ph.D. (1983) in Animal Science from Washington State University. He previously served on Farmland Industries' University Advisory Board and was an ex-officio member of the Board of Directors of the Dakota Feed Manufacturers.

Wayne P. Robarge, Ph.D. is Professor of Soil Physical Chemistry in the Department of Soil Science at North Carolina State University. His research interests include studies of emissions of ammonia from swine lagoons, temporal and spatial patterns in ambient ammonia and ammonium aerosol concentrations, nitrogen budgets using Geographical Information Systems (GIS), and dry deposition of ammonia and ammonium aerosols to crop and woodland canopies. He received his B.S. (1969) and M.S. (1971) from Cornell University, and Ph.D. (1975) in Soil Science from the University of Wisconsin-Madison. He currently serves on the USDA Task Force on Agricultural Air Quality. He currently conducts research from the North Carolina State University Animal and Poultry Waste Management Center as part of " An Integrated Study of the Emissions of Ammonia, Odor and Odorants, Pathogens and Related Contaminants from Potential Environmentally Superior Technologies for Swine Facilities."

Daniel A. Wubah, Ph.D. is associate dean College of Science and Mathematics at James Madison University. Prior to this, Wubah was chairperson of the Department of Biology at Towson University. His special expertise includes rumen microbiology and anaerobic zoosporic fungi. He received his B.S. and B.Ed. (1984) from the University of Cape Coast (Ghana) M.S. (1987) from the University of Akron, and Ph.D. from the University of Georgia. Wubah previously served on the NRC Panel for Review of Proposals Under the AID Research Grants Program for the Historically Black Colleges and Universities - Agriculture, Health, and Social Sciences. He is a member of the Board of Governors of the National Aquarium in Baltimore.

Kelly D. Zering, Ph.D. is associate professor Department of Agricultural and Resource Economics at North Carolina State University. His special expertise is economics of swine production and processing. He received his B.S. (1977) and M.S. (1980) from the University of Manitoba, and Ph.D. (1984) in agricultural economics from the University of California, Davis. Zering has extension responsibilities in the areas of swine management and marketing. He has completed research funded by EPA and the Animal and Poultry Waste

Management Center for "Economic Analysis of Alternative Manure Management Systems." He currently conducts research on manure technology evaluation funded by the North Carolina Attorney General—Smithfield Agreement via the Animal and Poultry Waste Management Center.

Ruihong Zhang, Ph.D. is associate professor Department of Biological and Agricultural Engineering at the University of California, Davis. Her main interests include control of gaseous and particulate emissions from animal feedlots, and wastewater treatment. She is a member of the USDA multi-state research project NCR-189, "Air Quality Issues Associated with Livestock Facilities" and a member of the American Society of Agricultural Engineers Committee on Environmental Air Quality. Zhang received her B.S. (1983) from InnerMongolia Engineering University (China), M.S. (1986) from the Northeast Agricultural University (China), and Ph.D. (1992) from the University of Illinois at Urbana-Champaign. She has a U.S. patent approved (filed by University of California, Davis) for a "Biogasification of Solid Wastes by Anaerobic Phased Solids Digester System."